SECOND EDITION

REAL Science Odyssey

Biology LEVEL 2

Teacher Guide

Blair H. Lee, MS

D1449350

Pandia PRESS

The publisher and author have made every attempt to state precautions and ensure all activities and labs described in this book are safe when conducted as instructed, but assume no responsibility for any damage to property or person caused or sustained while performing labs and activities in this or any RSO course. Parents and teachers should supervise all lab activities and take all necessary precautions to keep themselves, their children, and their students safe.

www.pandiapress.com

Biology LEVEL 2 Teacher Guide

Table of Contents

Table of Contents

RSO Biology 2 Teacher Guide
Introduction

RSO Biology 2 is a complete, yearlong biology course. The student textbook, student Workbook, and the Teacher Guide contain all the information you need to teach biology this year. This course was written so that it could be taught by all educators, even if you do not have a science degree. It is my goal to make biology accessible to all, educators and students alike.

Structuring a good science course is like building a house. A well-built house starts with a strong foundation on which all other floors rest. RSO Biology 2 starts with the fundamentals, and fact by fact builds from there, creating a strong foundation for future science knowledge to rest on.

While learning the fundamentals of biology, students need to learn and practice the methods used by biologists to establish the scientific facts and theories that form the basis of those fundamentals. Understanding the scientific method and application of the processes used by biologists guides students to an understanding of the open-ended thinking that is a part of science. The labs are closely paired with the written material; that way students can see how the results of experiments have led to a better understanding of how the living world works. The experiments also demonstrate the application of science principles. All of this together gives you a complete science course that teaches the core principles of biology while teaching the processes used to develop these principles.

Course Structure

RSO Biology 2 consists of three books: the student textbook, the student Workbook, and the Teacher Guide. If working with a group, several students can share a textbook, but I recommend each student have a copy of the Workbook for written assignments, lab findings, and assessments. The course is comprised of 32 chapters designed as a 36-week course, assuming each chapter is completed within one week plus some extra time for review and testing.

The Course Is Divided Into Seven Units

    I.   Organisms

    II.   Cells

    III.   Genetics

    IV.   Anatomy and Physiology

    V.   Evolution

    VI.   Ecology

    VII.   Classification

Each Chapter Has Sections

In the student textbook:

- Lesson

In the student Workbook:

- General biology labs and activities
- Optional* microscope labs (in most chapters)
- Famous Science Series (research assignment)
- Show What You Know (short answer and multiple choice sections)
- Optional* unit exams (found in the appendix)

   *Not optional for high school level

The Student Textbook

Students begin each chapter by reading a lesson in the textbook. Lessons are designed so that as students read, they are engaged through thought-provoking questions, and in many cases, by writing or coloring parts of diagrams found in the Workbook. This type of direct engagement when incorporated into learning material gives students ownership of the material.

   The following are the sections of each chapter found in the textbook and in the Workbook. The sections are represented by the RSO acronym R.E.A.L.—Read, Explore, Absorb, and Learn. They are presented here in the order you will find them in both books. Although you can switch the order of the general labs, microscope labs, and the Famous Science Series within a chapter, you should not change the order of the chapters themselves. Each chapter builds upon the prior one.

READ: The Lesson

In the textbook lesson the fundamentals of biology are explained and built on. These lessons are designed to get students thinking about key concepts, asking questions, and applying what they are learning to things they observe in the world around them. The lesson is at a reading level so that most students can read it independently. Science vocabulary and terminology are introduced in context. They are written in **bold italics**. Formal definitions can be found in the glossary of the textbook. Some lessons instruct students to color and label diagrams found in the Workbook.

The Student Workbook

After reading the chapter lesson in the textbook, students complete the labs, activities, research assignments, and assessments found in the Workbook for that chapter. Since students will be writing directly in the Workbook, it is recommended that each student has their own.

EXPLORE: General Labs

After reading the lesson, students turn to their Workbook to explore lesson concepts through experimentation. Students learn how scientists investigate and practice the scientific method in a meaningful way; gaining new insights into biology in the process. The general biology labs relate directly to the written lessons. The two are cohesive. Pairing lessons with labs that support the material studied is living science and is important to the understanding of science. Some labs have a math section. All math concepts are clearly explained with examples. Science is a good place to begin applying the abstract math concepts students have been learning all these years. Note that several labs require adult supervision.

EXPLORE: Microscope Labs

The microscope labs are optional, but to truly get the most out of this biology course, a microscope is highly recommended. Students completing Biology 2 for high school credit should complete the microscope labs. All living things are made of cells, and cells are really small, so small the you need a microscope to see them. There is something really special about the first time you see a cell up close, chloroplasts, the wing of a fly, or bacteria move. See the note about purchasing a microscope on page 9.

Some labs are conducive for formal lab write-ups, and some are not. I indicate those that are recommended in the student text.

ABSORB: Famous Science Series (FSS)

The skill of researching a topic is essential to being proficient at science. As the famous scientist Isaac Newton once stated, "If I have been able to see farther than others, it was because I stood on the shoulders of giants." By giants he means other great scientists. Scientists research what is known about a topic and build on that knowledge when making new discoveries.

The purpose of FSS is to sharpen the researching skills of students taking this course while they learn some interesting history relating to biology. Students are expected to research the questions in FSS on their own; the information is not found in the course material. How you have students conduct research is up to you. I feel that Internet research is adequate for FSS. But you might want to have your student do some library research as well.

There are 32 Famous Science Series topics in this course. The topics include famous scientists, famous pathogens, famous molecules, and famous scientific discoveries. Students will use FSS to learn topics more in depth as they relate to the lesson. If you want to reduce the amount of writing for students, you can have them orally report the results of their research to you.

LEARN: Show What You Know (SWYK)

Biology can be technical with a lot of new vocabulary words. I have tried to make the text as interesting as possible to keep students engaged in the material, but it is still important to have weekly assessments to ensure they have learned the key concepts. Show What You Know (SWYK) is the title of the question-and-answer section at the end of each chapter. I strongly recommend your students complete the SWYK assessments. They will help you assess whether your student understands the material being covered. If students do not do well in this section, you know they need to go back over the material before moving on to the next chapter. You can use this section like a written test, or you may choose to use SWYK as a format for open discussion. If students are taking the unit exams, this section will help immensely with their performance on the exams.

The Teacher's Guide

Your teacher guide is set up to resemble the student books. Each chapter section is reviewed in the teacher's guide with further explanation not found in the student text, as well as answers and suggestions. In addition, the teacher's guide contains the following:

Weekly Schedule

I have provided you with suggestions for scheduling each chapter based on teaching science two days, three days, and five days. The schedules will help ensure you complete the course in one school year.

Learning Goals

These are a list of all the important concepts in the chapter. Reviewing the learning goals can be particularly helpful when deciphering main ideas that shouldn't be missed from details that are nice, but not necessary.

Extracurricular Resources

This is a list of books and other resources that complement the material presented in the chapter. Links to website and online videos are provided at Pandia Weblinks (www.pandiapress.com/weblinks-biology2). Use these resources when your student's interest is sparked, or when you need further clarification on a concept.

Math this Week

The math concepts presented in a chapter are reviewed in the teacher guide. The math presented in a lab can be treated as optional, although I recommend that students at least attempt to complete it. Math is integral to a good science background. Math and science are intertwined in the same way spelling, punctuation, and grammar are to good writing.

Lesson Review

The lesson reviews included in this teacher guide are written as class notes. They can be used as notes or you can use them for question and discussion with your students. They are the main points from the lesson and are provided to assist you in teaching this course. When I taught as a community college professor, I would use a sheet of written notes, my lecture notes, as a guide to make sure I covered the important points and reviewed the material from previous lectures that related to the material being taught that day.

Unit Exam Answer Keys

The student Workbook contains the unit exams in its appendix. There are six of them. See page 12 of this teacher guide for information about administering and grading. The answer keys for the exams follow each unit in the teacher guide.

Grading

Grading is up to you, the instructor. Below are four possible grading schemes based on whether you are administering the unit exams and/or completing the microscope labs. There are grading scales provided for each unit exam, but you will have to determine the grade for each of the other parts of the course. The grading schemes below suggest how to weigh each part, if you choose to assign a grade. You, of course, are the teacher and will do what works best for you and your student.

1. Using all parts of the course
Unit exams = 40%
Microscope Labs = 15%
General Labs = 15%
FSS = 10%
SWYK = 20%

2. Not using unit exams
Microscope Labs = 20%
General Labs = 20%
FSS = 20%
SWYK = 40%

3. Not using microscope labs
Unit exams = 40%
General Labs = 20%
FSS = 15%
SWYK = 25%

4. Not using unit exams and microscope labs
General Labs = 30%
FSS = 30%
SWYK = 40%

Microscope

The microscope labs add a depth of understanding to any general biology course, but they can be left out in middle school. You could skip all the microscope labs and still have a high-quality middle-school-level biology course. Your students will need to use a microscope in high school, though, so you might want to think about that now. This way you will get a lot more use out of a microscope, and your student will be a microscope "expert" before high school.

If you are going to be purchasing a microscope, I suggest you invest in a nice one. This doesn't necessarily mean you have to spend your child's college fund, but you shouldn't waste your money on "toy" microscopes. For this course, and on into high school, you will need a compound light microscope. Compound microscopes have two lenses, the eyepiece and the objective lens, which work together to magnify the specimen. The type of compound light microscope used for these experiments is a bright field microscope. Bright field microscopes form a dark image against a more brightly lit background through the use of under lighting. Therefore you need a microscope that has an electric light. (I prefer direct current because it can be hard to tell when the battery is running down and this can affect the light coming from the base without you really noticing it.) Also, be sure to get one that has a fine focus knob.

You can use either a monocular microscope (one eyepiece) or a binocular microscope (two eyepieces). I give instructions for both types in the course. The advantage to the monocular scope is that it generally costs less, and it is easier to use for most, but not all, students. Some people have trouble focusing with both eyes open. The advantage of the binocular scope is that it magnifies up to 1000x with an oil immersion lens (the monocular scope magnifies up to 400x). There are a few labs where students get the opportunity to use the oil immersion lens if they have a binocular microscope. Either microscope is sufficient, however. The choice is entirely up to you.

Material List

Below is a list of items needed for the labs in each chapter. Refer to the student textbook for quantity and other details. The page # indicates the location of the lab in the student Workbook. The items marked with an asterisk * are those that are not readily available and need to be purchased through a science supply vendor. We recommend Home Science Tools (www.homesciencetools.com).

Chapter 1 Lab — p. 5
- ☒ Tape measure
- ☒ Graph paper
- ☒ Notebook paper
- ☒ Clipboard
- ☒ Calculator
- ☒ Outdoor area
- ☒ Field guides
- ☐ Plot markers

Chapter 1 MSLabs (1 & 2) — pp. 13 & 19
- ☐ Cutting mat or cardboard
- ☒ *Microscope
- ☒ White paper w/ black type
- ☒ Scissors
- ☒ X-Acto knife
- ☒ Tape
- ☒ *Microscope slide
- ☐ Catalogue with pictures

Chapter 2 Lab — p. 26
- ☒ Canning jars
- ☒ Soap & water
- ☐ Apples
- ☒ Knife
- ☒ Sugar
- ☒ Apple peeler
- ☒ Cutting board
- ☒ Tall pot w/ lid
- ☒ Cooking pot
- ☒ Plastic container
- ☒ Cooking source
- ☒ Wooden spoon
- ☒ Food processor
- ☒ Potato masher
- ☒ Permanent marker
- ☒ Timer

Chapter 2 MSLab — p. 29
- ☒ *Microscope
- ☒ *Slide w/ cover
- ☒ *Lens wipes
- ☒ Bottle cork
- ☒ X-Acto knife
- ☒ Cutting board
- ☒ Syringe
- ☒ Glass
- ☒ Water
- ☒ Tweezers

Chapter 3 Lab — p. 37
- ☒ White glue
- ☒ Super glue
- ☒ Plastic wrap
- ☒ Knife
- ☒ Scissors
- ☐ Toothpick
- ☒ Ruler
- ☒ Tape
- ☐ Toothpicks
- ☐ Clay (Sculpey)
- ☒ Cookie sheet
- ☒ Oven
- ☒ Bowl
- ☐ Plaster of Paris
- ☒ Container for mixing
- ☒ Stirrer
- ☒ Measuring cup
- ☒ Water

Chapter 3 MSLab — p. 43
- ☒ *Slides w/ covers
- ☒ *Methylene blue stain
- ☒ *Oil, for oil immersion
- ☒ Cleaner for oil
- ☒ Plastic spoon
- ☐ Yellow onion
- ☒ X-Acto knife
- ☒ Paper towel
- ☒ Syringe
- ☒ Water
- ☒ *Microscope

Chapter 4 Lab — p. 53
- ☐ Worksheets from Web
- ☒ Internet access
- ☐ Food items
- ☒ Printer

Chapter 4 MSLab — p. 61
- ☐ Labels from flour bags
- ☐ Whole-wheat flour
- ☐ White flour
- ☒ Teaspoon
- ☒ Butter knife
- ☒ Colored pencils
- ☒ *Slides w/ covers
- ☒ Syringe
- ☒ Water
- ☒ Toothpicks
- ☒ Glass bowls
- ☒ *Microscope
- ☒ Iodine

Chapter 5 Lab — p. 71
- ☒ Measuring cup
- ☒ Tablespoon
- ☒ Teaspoon
- ☒ Cornstarch
- ☒ Iodine
- ☒ Water
- ☒ Zipper-lock plastic bag
- ☒ Wide glass
- ☒ Paper

Chapter 5 MSLab — p. 75
- ☒ *Microscope
- ☒ Flashlight or desk lamp
- ☒ A helper
- ☐ Kernel corn
- ☒ Tweezers
- ☒ X-Acto knife
- ☒ Cutting board
- ☒ Water
- ☒ *Slide
- ☒ Iodine
- ☒ Small dish

Chapter 6 Lab — p. 81
- ☒ Colored pencils
- ☒ Sunny day
- ☐ Fruit or vegetable snack
- ☒ Glue and construction paper (optional)

Chapter 6 MSLab — p. 86
- ☐ One leaf from a thick plant
- ☒ X-Acto knife
- ☒ *Microscope
- ☒ *Slide w/ cover
- ☒ Water
- ☒ Syringe
- ☒ Green and gray pencil
- ☒ *Oil, for oil immersion
- ☒ Cleaner for oil

Chapter 7 Lab — p. 95
- ☐ Mini marshmallows
- ☒ Beads
- ☒ Pipe cleaners
- ☐ Large marshmallows
- ☒ Skewer
- ☐ Toothpicks
- ☒ Scissors

Chapter 7 MSLab — p. 99
- ☐ Sports drink
- ☒ Timer
- ☒ Rubbing alcohol
- ☒ Meat tenderizer
- ☐ Toothpick
- ☒ *Test tube
- ☒ *Pipette
- ☒ *Slide w/cover
- ☒ *Methylene blue stain
- ☒ Water
- ☒ *Microscope
- ☒ Dish soap
- ☒ Cup or glass

Chapter 8 Lab — p. 107
- ☐ Poster board
- ☒ Pipe cleaners
- ☐ Mini marshmallows
- ☒ Markers
- ☒ Ruler
- ☒ Computer & printer
- ☒ CD
- ☒ Glue
- ☒ Can
- ☒ Cup
- ☒ Yarn
- ☒ Beads

Chapter 8 MSLab — p. 112
- ☒ *Microscope
- ☒ *Prepared slide of an allium (onion) root tip

Chapter 9 Lab — p. 119
- ☒ Colored pens or pencils
- ☒ Scissors
- ☒ Stapler

Chapter 9 MSLab — p. 125
- ☒ *Microscope
- ☒ *Prepared slide of a *Lilium* (Lily), anther meiosis

Chapter 10 Lab & Act. — pp. 132 & 142
- ☒ Family members
- ☒ Colored pencils or markers
- ☒ Scissors
- ☒ Coin

Chapter 10 MSLab — p. 138
- ☒ Hair strands
- ☒ *Slides
- ☒ *Slide covers
- ☒ Scissors
- ☒ Tape
- ☒ Syringe
- ☒ Water
- ☒ *Microscope

Chapter 11 Dissection Lab — p. 155
- ☒ *Preserved frog
- ☒ Safety goggles
- ☒ *Dissecting pins
- ☒ *Dissecting tray
- ☒ Paper towels
- ☒ Gloves
- ☒ *Forceps
- ☒ Tape measure
- ☒ Scissors
- ☒ *Slides
- ☒ *Slide covers
- ☒ Tweezers
- ☒ Medicine dropper
- ☒ Baggies
- ☒ X-Acto knife

Chapter 12 Dissection/MSLab — p. 169
- ☒ Plant
- ☒ X-Acto knife
- ☒ Colored pencils
- ☒ *Microscope
- ☒ *Slides w/ covers
- ☒ Syringe
- ☒ *Methylene blue

Chapter 13 Dissection Lab — p. 179
- ☒ Flower
- ☒ Lima bean
- ☒ X-Acto knife
- ☒ Scissors
- ☒ Tape
- ☒ Magnifying glass

Chapter 14 Labs 1 & 2 — pp. 189 & 193
- ☐ Lemon
- ☒ Wire
- ☒ Nails or screws
- ☒ Pennies
- ☒ Calculator
- ☒ Salt
- ☒ Knife
- ☒ Paper towels
- ☒ Rubber band
- ☒ Stringed Instrument

Chapter 15 Lab 1 — p. 202
- ☐ Corrugated cardboard
- ☒ Nail
- ☒ Blindfold

Chapter 15 MSLab — p. 204
- ☒ *Microscope
- ☒ *Slide w/cover
- ☒ Water
- ☒ *Methylene blue
- ☒ Tissue
- ☒ Butter knife

Chapter 15 Lab 2 — p. 206
- ☐ 2 liter bottle
- ☒ Sink
- ☒ Scissors
- ☒ Coffee filters
- ☒ Gravel
- ☒ Sand
- ☒ Cotton balls
- ☒ "Dirty" water
- ☒ Camera or colored pencils

Chapter 16 Labs 1 & 2 — pp. 214 & 221
- ☒ Room w/ thermostat
- ☒ Flashlight
- ☒ Test subject
- ☒ Watch or timer
- ☒ Family member
- ☒ Paper & pen

Chapter 17 MSLab & Lab 1 — pp. 228 & 229
- ☒ *Slide
- ☒ *Slide cover
- ☒ Needle or pin
- ☒ Rubbing alcohol
- ☒ Soap & water
- ☒ *Microscope
- ☒ Water bottles
- ☒ Scissors
- ☒ Duct tape
- ☒ Timer
- ☐ Vinyl tubing
- ☒ X-Acto knife

Chapter 17 Lab 2 — p. 235
- ☒ Metric measuring stick or tape
- ☒ Balloon
- ☒ Calculator
- ☒ Another person

Chapter 18 Lab & Dissection/MSLab — pp. 245 & 250
- ☒ House or car
- ☒ Book
- ☐ Chicken wing
- ☒ Paper towels
- ☒ *Microscope
- ☒ *Slide w/cover
- ☒ Syringe
- ☒ Methylene blue
- ☒ Gloves
- ☒ Scalpel or knife
- ☒ Scissors
- ☒ Cutting board
- ☒ Freezer
- ☒ Desk lamp

Chapter 19 Labs 1 & 2 — pp. 264 & 272
- ☒ Paper
- ☒ Calculator
- ☒ Another person
- ☒ Outdoor area

Chapter 20 Lab — p. 277
- ☒ Large work space
- ☒ Tape measure
- ☒ Marker
- ☒ Cardboard or card stock
- ☐ Poker chips or checkers
- ☒ Glue
- ☒ Scissors
- ☒ Colored pencils
- ☒ Roll of banner paper (optional)

Chapter 20 MSLab — p. 287
- ☒ *Microscope
- ☒ *Slide
- ☐ Dead winged insect specimens
- ☒ Flashlight
- ☒ Scalpel or knife

Chapter 21 Lab — p. 295
- ☐ Pompoms—brown, black, gray, and white
- ☒ Black magic marker
- ☒ Another person
- ☒ Timer
- ☒ Inside carpeted area

Chapter 21 MSLab — p. 301
- ☒ Animal hair and fur samples
- ☒ *Microscope
- ☒ *Slides w/ covers
- ☒ Water
- ☒ Dropper

Chapter 22 Lab — p. 309
- ☒ Pie or cake pan
- ☐ Clay
- ☒ Cooking spray
- ☒ Items for cast impressions—shells, bones, leaves, rocks, etc.
- ☐ Plaster of Paris
- ☒ Mixing container
- ☒ Measuring cup
- ☒ Water
- ☒ Towel
- ☒ Stir stick

Chapter 22 MSLab — p. 312
- ☒ *Microscope
- ☒ Flashlight
- ☐ Sedimentary rock
- ☒ Paper
- ☒ Pencil w/eraser
- ☒ Magnifying glass (optional)

Chapter 23 Lab — p. 319
- ☒ Scissors
- ☒ Glue
- ☒ Large sheet construction paper (optional)
- ☒ Internet connection

Chapter 23 MSLab — p. 326
- ☒ *Microscope
- ☒ Magnifying glass
- ☒ Slice of wood
- ☒ Sandpaper
- ☒ Top lighting
- ☒ Paper

Chapter 24 Lab & Act. — pp. 333 & 336
- ☒ Internet access
- ☒ World map
- ☒ Shoebox
- ☒ Drawing paper
- ☒ Art supplies, glue, scissors
- ☒ Nature magazines
- ☐ Various supplies to make biome diorama

Chapter 24 MSLab — p. 341
- ☒ *Microscope
- ☒ Soil samples
- ☒ Top lighting
- ☒ *Slides
- ☒ Spatula

Chapter 25 Lab & MSLab — pp. 347 & 350
- ☒ Colored pens or pencils
- ☒ Field guides (optional)
- ☒ *Microscope
- ☒ *Slides
- ☒ *Slide cover
- ☒ Freshly picked grass
- ☒ Pliers
- ☒ Water

Chapter 26 Lab — p. 357
- ☒ Water
- ☒ Airtight container
- ☒ Pea gravel or pebbles
- ☐ Activated charcoal
- ☐ Spanish moss
- ☒ Soil
- ☒ Plants

Chapter 26 MSLab — p. 360
- ☒ *Microscope
- ☒ *Slides
- ☒ *Slide covers
- ☒ Water
- ☐ *Legume inoculant (rhizobacteria)

Chapter 27 Lab — p. 365
- ☒ Potted plants
- ☒ Dishes
- ☒ Jars w/ lids
- ☒ Tablespoon
- ☒ Marking pen
- ☒ Potting soil
- ☒ White vinegar
- ☒ Marking tags
- ☒ Distilled water
- ☒ Measuring cup
- ☒ Camera (optional)

Chapter 27 MSLab — p. 370
- ☒ Nonmetallic bowls
- ☒ Distilled water
- ☒ White vinegar
- ☒ Leaves
- ☒ Measuring cup
- ☒ X-Acto knife
- ☒ *Slides
- ☒ *Microscope
- ☒ *Slide covers
- ☒ Water
- ☒ Syringe

Chapter 28 Lab & Research — pp. 377 & 386
- ☒ Scissors
- ☒ Glue or tape
- ☒ Friends
- ☒ Magnifying glass
- ☒ Envelopes
- ☒ X-Acto knife
- ☒ Leaves
- ☒ Computer & printer
- ☒ Internet/library access

Chapter 29 MSLab — p. 395
- ☐ Yogurt
- ☒ Toothpick
- ☒ Water
- ☒ Dropper
- ☒ *Microscope
- ☒ *Slides
- ☒ *Slide covers
- ☒ *Methylene blue

Chapter 30 Lab & MSLab — pp. 403 & 405
- ☒ Field guides (optional)
- ☒ Blade of grass
- ☒ Leaf
- ☒ Scalpel or paring knife
- ☒ *Slides
- ☒ *Slide covers
- ☒ Water
- ☒ Syringe
- ☒ *Microscope

Chapter 31 Lab/MSLab — p. 411
- ☒ Magnifying glass
- ☐ Arachnid specimen
- ☐ Insect specimen
- ☒ *Microscope
- ☒ Liquid paper
- ☒ Scalpel
- ☒ *Slides
- ☒ *Slide covers
- ☒ Copy paper
- ☒ Syringe
- ☒ Water
- ☒ Tweezers
- ☒ Flashlight

Chapter 32 Lab & Dissection/MSLab — pp. 421 & 425
- ☐ Banana
- ☒ Baggies
- ☒ Yeast
- ☒ String
- ☒ Ruler
- ☐ Mushroom
- ☒ Cutting board
- ☒ Flashlight
- ☒ Scalpel
- ☒ *Slides
- ☒ *Microscope
- ☒ *Methylene blue
- ☒ Syringe
- ☒ *Forceps

RSO Biology 2 Teacher Guide
Introduction to Student Unit Exams

There are six exams spanning seven units. The exams are found in the appendix of the student text. Answers to the exams are found in this teacher guide following each unit beginning with Unit 2. The exams have multiple-choice questions, vocabulary matching, true/false sections, and short written answers. There is no cumulative mid-term or final exam. One could be made by combining questions from the unit exams.

What Each Exam Covers

Exam 1: Units I Organisms and Unit II Cells, Chapters 1 – 6
Exam 2: Unit III Genetics, Chapters 7 – 10
Exam 3: Unit IV Anatomy and Physiology, Chapters 11 – 19
Exam 4: Unit V Evolution, Chapters 20 – 23
Exam 5: Unit VI Ecology, Chapters 24 -27
Exam 6: Unit VII Classification, Chapters 28 – 32

As the instructor, it is up to you how the exam is administered.

Possible Options

1. A closed-book exam with no notes
2. A closed-book exam with one sheet of notes (more pages of notes than this just get in the way)
3. An open-book exam
4. Don't use it as a exam at all; use it as a review

Structure of the Exams

- Multiple choice questions
- Vocabulary match
- Short answer questions
- The material from the Famous Science Series and the labs is not tested.
- Each exam is 100 points. Most of the exams have opportunities for extra credit.

Grading the Exams

- The questions that require written answers can make grading a little more difficult. Use my answers as a guide. Just remember, partial credit should be applied to these questions if students get most, but not all, of the answer correct.
- Students can get more than 100 percent on the exam if they get the extra credit points.

After the Exam

- Go over with your students any questions they missed. Use mistakes as an opportunity to learn.
- You can hand back the exam with incorrect answers marked and give half credit for any exam questions students correct. I like to do this because it keeps the focus on the primary reason for studying a course: to learn the material, NOT to get a grade.

Teacher Guide

Unit I: Organisms

Chapter 1: All Living Things

WEEKLY SCHEDULE

Two Days

Day 1
- ❑ Lesson
- ❑ Read Poem
- ❑ Lab

Day 2
- ❑ MSLabs
- ❑ FSS
- ❑ Lesson Review
- ❑ SWYK

Three Days

Day 1
- ❑ Lesson
- ❑ Read Poem

Day 2
- ❑ Lab

Day 3
- ❑ MSLabs
- ❑ FSS
- ❑ Lesson Review
- ❑ SWYK

Five Days

Day 1
- ❑ Lesson

Day 2
- ❑ Read Poem
- ❑ Lab

Day 3
- ❑ MSLabs

Day 4
- ❑ FSS

Day 5
- ❑ Lesson Review
- ❑ SWYK

FSS: Famous Science Series
MSLab: Microscope Lab
SWYK: Show What You Know

Introduction

Unit I is an introductory unit that is one chapter long. There are seven units in this book. Unit I is the only unit that does not have a separate exam available. Unit exam 1 covers elements from Units I and II.

Preparation for Lab 1: Locate an area for the plot study experiment.

Learning Goals

- Learn the nine characteristics that define organisms as living.
- Study about viruses and their big effect on all organisms.
- Understand how viruses reproduce.
- Learn the parts of a microscope.
- Learn how to focus a microscope.
- Investigate how plot studies are conducted.
- Find out about the numbers and species of wild plant and animal life near where you live.

Extracurricular Resources

Books

Five Kids & A Monkey Investigate a Vicious Virus, Blair, Beth L. and Riccio, Nina

Epidemic! The Battle against Polio, True Peters, Stephanie.

Jonas Salk, (Inventors and Creators Series), Durrett, Deanne

Jonas Salk and the Polio Vaccine, Krohn, Katherine. This book is written in graphic format.

Small Steps: The Year I Got Polio, Kehret, Peg

Jonas Salk: Conquering Polio (Lerner Biographies), Sammartino McPherson, Stephanie

Jonas Salk: Creator of the Polio Vaccine (Great Minds of Science), Tocci, Salvatore

Jonas Salk and the Polio Vaccine (Unlocking the Secrets of Science), Bankston, John

West Nile Virus (Diseases and Disorders), Abramovitz, Melissa

West Nile Virus: Epidemics Deadly Diseases Throughout History, Margulies, Phillip

Ebola Virus (Diseases and People), Willett, Edward

Understanding Viruses with Max Axiom, Biskup, Agniesezka

Killer Virus (Choose Your Own Adventure(R), Montgomery, R.A.

Franklin Delano Roosevelt: Champion of Freedom, Kudlinski, Kathleen

Online

Visit Pandia Weblinks for videos and websites recommended for this chapter:

www.pandiapress.com/weblinks-biology2

Pandia PRESS

Lesson

What Is Living?

The lesson for this week explains the nine characteristics that define life. The poem reinforces these characteristics. Students at this level are competent at knowing when something is alive and when it is not. So competent, in fact, that they might gloss over this section. That is why I introduce the intriguing and thought-provoking example of viruses. The debate about viruses is a real-world application of the characteristics defining life. As students think through the debate about how viruses should be categorized, living or not, they will have to think through the characteristics used to define an organism as living. This will help reinforce these defining characteristics. Students are asked to come to their own conclusion about whether viruses should be reclassified as organisms. There is no right or wrong answer, in my opinion. I am a pragmatist, though. Viruses are what they are. The argument is really just a matter of definition, but definitions are very important when classifying organisms, so maybe the definition is not so trivial after all.

Students are asked to come to their own conclusion as to whether viruses should be reclassified as organisms. This can be done briefly on paper or orally. I would consider all well-thought out answers as correct. If you want a topic to debate over dinner, this would be a good one.

Lab

What's Out There? Plot Study

This is the type of lab I would expect to see at the start of any general biology class. Biology is the study of life. I think all biology students should start with a study of the organisms near where they live. A plot study is a great way to do that. An interesting addition to this lab would be to revisit the plot two or three times over the year and check out the seasonal changes at the plot.

You need to give some thought to the location of the plot. Choose somewhere that you can sit and observe for a while. Be prepared to help with the measuring of the plot lines, counting organisms, and drawing the plot. Do not let the calculations page seem overwhelming. There are a lot of words because I am trying to walk students through the process. Just worry about counting animals and filling in the data table when you are in the field.

If your plot size varies, the size of your rectangle will be different. For example, if your plot is 1m x 2m, then draw a block that is 10 squares by 20 squares on the graph paper. And for lab calculations, the area of this plot would be $2m^2$ (2m x 1m).

This is a good lab for writing a report lab. On the opposite page is an example of a completed lab report and data tables. Your students' reports will vary, and should include a drawing of the plot.

Math This Week

1. Measuring and marking the perimeter of the plot.
2. Mapping the area of the plot—this is a great activity to increase spatial awareness.
3. Drawing the mapped plot to scale.
4. Calculating the area of the plot.
5. Types of problem solving are: estimating, going from percent to decimal, multiplying, dividing, adding, and rounding.

Chapter 1: Lab Report

Name: _____   Date: _____

Title/Location: _Plot Study From an Irrigation Ditch in the Eastern Sierras_

Hypothesis

Using a small plot study of an irrigation ditch, I think I will get a good estimate of the plant and animal species for a larger area.

Procedure

I conducted a plot study on an irrigation ditch in the Eastern Sierras. I mapped the organisms on a 2m x 2m plot. I used the results from this study to estimate the number of different species and over-all number for each species that would be in 100m². I measured my plot along the bank of the creek. I began at one corner and methodically drew what was within the plot boundaries. I used the field guides to identify plants and animals. I recorded and counted the different animal and plant species I found onto my data tables. I used these numbers to calculate an estimate of the number and variety of plants and animals for a larger, 100 m², area.

Observations

· The flow of water is stronger in the middle of the ditch than on the sides. Most plants and animals prefer either the stronger flow or the weaker flow, but not both.
· The spider spun a web over the water. It hid on the side of the ditch until an insect landed on the web. Then it ran out to catch its dinner.
· In the past, I have seen an occasional fish in the ditch, but not today.
· Dragonflies flew over the ditch and landed on the grass out of the water but never in the water.
· Because it has been very cold this spring, I did not see any mosquito larvae.

Results and Calculations

I estimated the number for each species of animals and plants that I found (see attached Data Tables 1 and 2). On Table 3, I estimate there are nine animal species in 100m2 area of the ditch, and I estimate there are six plant species in 100 m2 area of the ditch. Notes: * I did not observe any fungi, so I did not include that in my Lab Report.
* Algae are plant-like protists, not plants, but I listed them under "Plants" in the table.
* For the Tables and Calculations, see attached

Conclusions

In conclusion, I think I got a good estimate of the number of plant and animal species in the irrigation ditch. I also think this was a good method for estimating the over-all number for each species.

Lab Calculations

Tables 1 and 2

How many of each type of organism is in a 100 square meter area, 100m²?

A. Calculate the area of my plot.

   2m x 2m = 4m² plot

B. Calculate how many of these plots would fit into 100m2.

   100m2 ÷ 4 =

25 of my plots would fit into 100m2

Table 3

I estimate that 10% were missed.
I turned 10% into a decimal: 10 ÷ 100 = .1
8 animal species x .1 = .8
.8 is rounded up to 1
How many animal species, would you expect to find in 100m²?
8 + 1 = 9
5 plant species x .1 = .5
.5 is rounded up to 1
How many animal species, would you expect to find in 100m²?
5 + 1 = 6

Chapter 1: Data Tables

Table 1

Animals (list each species)	# of species in my plot	# of my plots that fit into 100m²	Estimate # of species for 100m²
water strider	14	25	350
small snail	20	25	500
large snail	2	25	50
small crickets	5	25	125
little swimming black insects (probably larvae)	lots (more than 100)	25	2500+
white swimming insects	11	25	275
spider	1	25	25
sylphid beetle	1	25	25

Total animal species= 8

Table 2

Plants (list each species)	# of species in my plot	# of my plots that fit into 100m²	Estimate # of species for 100m²
algae, listed as plant	10 cm x 10 cm	25	2500 cm²
long grass growing in water at edge of ditch	10 clumps	25	250 clumps
tall grass with seeds in ditch	3 clumps	25	75 clumps
short green grass growing under water	17 clumps	25	425 clumps
plant w/ spiky leaves	1	25	25 clumps

Total plant species= 5

Table 3

	# Species found in my plot	Estimate # missed	Estimate # of species in 100m²
Animals	8	1	9
Plants	5	1	6

Microscope Labs 1 and 2

Your Microscope: Parts and Focus

This lab is a beginner microscope lab. Students will learn the names for the parts of the microscope as well as how to focus the microscope. If your students are experienced microscope users, you might find they do not need to do this lab. Through the remainder of this course, it will be assumed your student has performed this lab, and knows the terminology and how to use a microscope.

The instructions for the first few microscope labs are very long. They are written for those students who have never used a microscope and want to know it all. They are instruction manuals detailing proper technique, procedure, and terminology. Later in the text, the instructions for the microscope labs shorten considerably. Through continued use, students will become very good at using a microscope.

Part 1: Parts of the Microscope

The purpose of the first part of this lab is to teach students the names of all the relevant parts of the microscope. See the next page for correct labeling.

Students are not expected to memorize microscope terms. I expect they will learn the names of the parts of the microscope through continued use.

Part 2: How to Focus Your Microscope

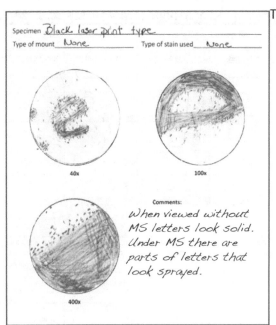

This section instructs students on the correct procedure for focusing a microscope. Students are expected to draw three pictures, each is a microscopic view of the paper and ink at one of the three magnifications. The oil immersion lens will NOT be used with this lab. Next, students will look at a color picture from a catalogue. It is really interesting to see how all the colors in a catalogue are made by combining the same four colors of dots in different proportions.

This is an example of a completed microscope view sheet. Throughout this guide you will find several such examples. These are for reference only. What your student views through the microscope could be vastly different than what is pictured in these examples.

Lab sheet suggested answers:

"K" indicates black in the four-color print processing used for your catalogue picture (CMYK). But it doesn't stand for the word "black." Research what it does stand for and why. *K stands for the word "key." Way back when color printing started, the black plate was called the key plate because it contained the artistic detail or "key" information. Today we say that K means "black," so as not to confuse it with B (blue) in the RGB color model.*

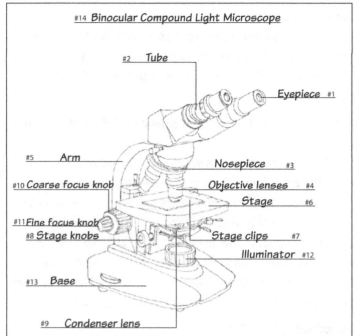

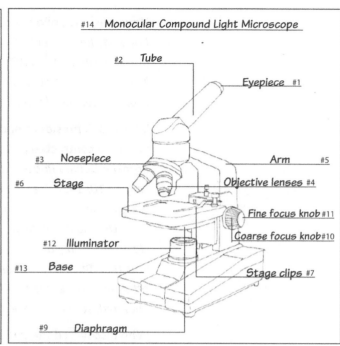

Famous Science Series

Polio

Living or not, viruses have a big effect on organisms. Most children in this country are vaccinated against polio, so children today might not have heard of it. There was a time when people lived in fear of it. The vaccine for polio was invented by a true American hero, Jonas Salk. Salk spent his life trying to find cures for deadly diseases like AIDS and polio. When he did find a vaccine that worked, he refused to patent it. Vaccines that are not patented are cheaper and therefore available to more people.

The following example is very detailed. Students are not expected to answer with as much detail. It is simply for your and their information.

What is polio? How is it transmitted?
Poliomyelitis or polio is a virus that infects people. Polio is transmitted through polio-contaminated feces. The route of transmission is usually from a person's hands to their mouth. You can also get polio from sharing eating utensils with an infected person.

What does it do to a person who is infected with it? What is paralytic polio?
Ninety percent of the people who get polio recover from it with no ill effects. The other ten percent develop symptoms. One percent of these people develop paralytic polio. Paralytic polio causes paralysis. This in turn can lead to deformities of the hips, ankles, and/or feet. Polio can also cause breathing problems. People who suffered from these breathing problems sometimes had to use an iron lung to help them breathe in order to stay alive. In severe cases, people infected with polio died.

How long has polio been infecting people?

Polio has been infecting people for thousands of years. A stone carving from Egypt dated to about 1500 BCE shows a boy with shrunken legs caused by the virus. Tiny Tim in Charles Dickens's A Christmas Carol *was probably a victim of polio. Polio mainly infects children.*

Which U.S. president had polio? When did he serve as president? How old was he when he contracted polio?

Franklin Delano Roosevelt, FDR, was 39 years old when he contracted polio on August 10, 1921. He was the 32nd president. "Once you've spent two years trying to wiggle one toe, everything is in proportion," Franklin Delano Roosevelt said in 1945.

FDR had paralytic polio, which caused him to be paralyzed from the waist down. He was the only disabled president. He served as president from 1933 to 1945. He was the only U.S. president to serve three terms. He died less than three months after he was elected to his fourth term. The United States Constitution has since been changed, so that no one can be elected for more than two terms as president.

Who discovered the polio vaccine?

Polio used to be widespread until Dr. Jonas Salk, a true American hero, discovered the polio vaccine. The polio vaccine was made available to the public in 1955. He did not patent his polio vaccine discovery, because it would have drastically increased the medicine's price. He freely distributed the polio vaccine so every child could be saved from contracting this potentially crippling disease. In addition to polio, Dr. Salk dedicated his life to researching the causes, preventions, and cures of influenza, cancer, and AIDS.

Show What You Know

All Living Things

Answers:

This penguin is a living being. It is a(n) *organism*.

The penguin eats fish. This is how it takes in *energy*.

After it eats fish, it has to get rid of *waste*.

Laying eggs is part of how the penguin *reproduces*.

Penguins *move* when they swim through the water.

Penguins ruffle up their feathers, trapping warm air near their bodies to help them stay warm. This is one way penguins *respond to their environment*.

This penguin's blood carries food to its cells and waste away from its cells. That is because penguins have *circulation*.

This penguin is made from many more than one *cell*.

Penguins get energy from the food they eat. Penguins have *respiration*.

A baby penguin *grows* after it hatches from the egg on its way to becoming an adult.

Lesson Review

All Living Things

living beings = organisms

The nine characteristics that define ALL organisms:
1. All organisms are made of one or more cells.
2. All organisms take in energy.
3. All organisms get rid of waste.
4. All organisms move.
5. All organisms grow.
6. All organisms reproduce.
7. All organisms respond to their environment.
8. All organisms have some type of circulation.
9. All organisms have some type of respiration. *There are two types of respiration, aerobic (with oxygen) and anaerobic (without oxygen). The example used in this chapter is aerobic respiration. Both types of respiration will be covered in more detail in chapter 6.

If something does not have all nine of the characteristics, it is not defined as living. Viruses reproduce and possibly respond to their environment. Most scientists do not define viruses as organisms because they do not have the other seven characteristics needed to define life.

Viruses reproduce by attaching to a cell and injecting parts of itself into the cell. These parts turn the cell into a virus-making factory.

Unit II: Cells

Chapter 2: Types

WEEKLY SCHEDULE

Two Days

Day 1
- ❑ Lesson
- ❑ Lab
- ❑ FSS

Day 2
- ❑ MSLab
- ❑ Lesson Review
- ❑ SWYK

Three Days

Day 1
- ❑ Lesson
- ❑ Lab

Day 2
- ❑ MSLab

Day 3
- ❑ FSS
- ❑ Lesson Review
- ❑ SWYK

Five Days

Day 1
- ❑ Lesson

Day 2
- ❑ Lab

Day 3
- ❑ MSLab

Day 4
- ❑ FSS

Day 5
- ❑ Lesson Review
- ❑ SWYK

Introduction Unit II

Unit II consists of five chapters and it covers the topic of cell biology. All organisms are made from one or more cells. Therefore, an understanding of the structure and function of cells is fundamental to an understanding of biology. Cells and the molecules that build them cannot be seen individually without the help of an optical device, like a microscope. Three of the five microscope labs in Unit II examine cells and their parts. The other two microscope labs look at molecules that make cells. In Unit II, several of the labs, Famous Science Series topics, and activities are related to health issues. We are made of cells; how healthy we are is directly related to how well we take care of our cells.

Introduction Chapter 2

Chapter 2 explains the cell theory and the three things all cells share in common. It also classifies the two main groupings for cell types.

The history of science is filled with colorful characters and important discoveries. The reasons for, or methods used, when making discoveries are sometimes pretty strange. The two labs and one of the discoveries made by the scientist chosen for the Famous Science Series highlight this.

Preparation for microscope lab: The day before, put the cork in a glass of water. It will float, but it will absorb some of the water. Doing this makes it easier to get a thin slice of cork.

Learning Goals

- Memorize the three parts of the cell theory.

- Identify the three components all cells have.

- Understand the basic difference between eukaryotic and prokaryotic cells.

- Learn the technique for making wet mount slides.

- Learn about the history and process of canning and examine some basic food safety principles as they relate to canning.

- Perform the historic experiment when cells were discovered and named.

- Research the history of the first scientist to see living cells with a microscope.

FSS: Famous Science Series
MSLab: Microscope Lab
SWYK: Show What You Know

Extracurricular Resources

Books

The Basics of Cell Life with Max Axiom, Super Scientist, Keyser, Amber J.

Enjoy Your Cells, Balkwill, Fran

Germ Hunter: A Story About Louis Pasteur, Alphin, Elaine Marie

Louis Pasteur, Spengler, Kremena

Pasteur's Fight Against Microbes, Birch, Beverly

Louis Pasteur: Founder of Modern Medicine, Tiner, John Hudson

Robert Hooke: Natural Philosopher and Scientific, Burgan, Michael

Micrographia–Some Physiological Descriptions of Minute Bodies Made by Magnifying Glasses with Observations and Inquiries Thereupon, Hooke, Robert

Online

Visit Pandia Weblinks for videos and websites recommended for this chapter:

www.pandiapress.com/weblinks-biology2

Lesson

You Are a Eukaryote

The lesson for this week defines the cell and its role as the building block for all living organisms. The cell theory is explained. This theory is one of the central tenets of biology. There is an explanation of the three components shared by all cells.

The initial classification used for organisms is on the cellular level. It is based on where the genetic material is in the cell. The members of domain Bacteria and domain Archaea have prokaryotic cells without a nucleus. The members of domain Eukarya have eukaryotic cells with a nucleus. (The domain level has been added to the top of the classification system in taxonomy. For an explanation, please refer to chapter 28.)

Lab

Death to the Prokaryotes!

In this course I have, whenever possible, paired labs with theory. I feel strongly that a good science text has labs that directly relate to the theory sections. The theory sections are the written lessons. In addition, for this middle school text, I wanted some of the labs, activities, and Famous Science Series to focus on pertinent health concepts.

Math This Week

Food Canning
1. Measuring volume.
2. Diving into parts.

   As I was developing a lab for this chapter, two food safety scares occurred in the same month. One was caused by unsanitary practices at a food manufacturing plant involving peanut butter, and the second was an E. coli scare affecting spinach. Both of these events resulted in illnesses caused by unicellular prokaryotic pathogens. It is important to include a food safety lab in this course, and apropos when teaching prokaryotes. I want students to understand the potential devastation something microscopic can cause if we aren't careful with the food we eat.

   Most people take it for granted that the food we eat is safe. Canning is an important method of preserving food. In this lab, students will learn how food is canned and how harmful pathogens in food are killed. They will also learn the interesting history of canning. The process of canning food was invented so that Napoleon could better feed his troops during wartime.

   Students will make applesauce in this lab. They will divide it into four parts, one part is to be eaten right away, but one part will be left sitting out. This part will go bad because it is not processed. It should NOT be eaten and will be thrown away. You will know when it has spoiled. The other two parts will be processed. One of these will be eaten in two weeks, the other in two months. The canned applesauce would be good for longer than that, but there needs to be an end date for this experiment.

   Usually when a person cans food, they can a larger quantity than that canned for this experiment. The amount to be canned is kept small for those students and teachers who do not want to peel a bushel of apples. If you want to increase the amount, just be consistent with the amounts of the ingredients.

Lab Sheet Suggested Answers

If old rotting apples were used . . . *Rotting is caused by bacteria and other microorganisms. The more bacteria on the fruit, the harder it is to get rid of. The whole point is to eat non-rotten food.*

If bruises were not cut from the apples . . . *Bruised apples are damaged. Bruised sites can have small tears in them, which increases the threat that bacteria might have gotten inside.*

If the jars, rims, and lids were not clean . . . *You are cleaning away things that might spoil your food and make it unsafe to eat.*

If the seal between the jar and rim was not tight . . . *Microorganisms can get into the jars and spoil the food.*

If the applesauce was not cooked as long as it should have been . . . *You might not have killed all the microorganisms.*

Microscope Lab

Discovering Cells

Yikes! Another microscope lab with lots of explanation and a long procedure section. These beginning microscope labs are written to give you, the teacher, more options. You can spend a lot or a little time on them. This lab gives step-by-step instructions in how to assemble and view a wet mount slide. The microscope labs become much shorter once the various parts and procedures we use in this book have been explained.

This lab recreates the historically important lab in which Robert Hooke documented the first microscopic view of a cell. The cells Hooke saw were cork cells. Cork is made from dead tree bark. Unlike cells that are alive, dead cells are "empty." Students will only see the cell walls of cork. During this lab, students will learn the important microscope technique of making wet mount slides.

Microscope Note: If the cork is sliced too thick, you will not have a clear view of each cell.

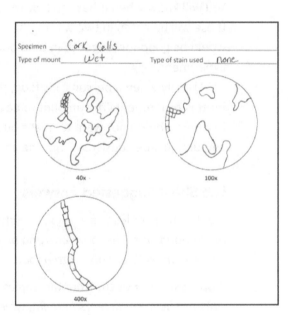

Famous Science Series

Antonie van Leeuwenhoek

Suggested Answers

Why is Antonie van Leeuwenhoek (LAYU-wen-hook) famous? What did he discover? What did he use to discover them? *Antonie van Leeuwenhoek has been called the father of microbiology. He is best known for his work to improve microscope technology. He became fascinated with how lenses magnify. He began grinding and polishing his own lenses. At the time, his lenses were some of the finest made.*

With his microscope, van Leeuwenhoek was the first person to see bacteria, unicellular eukaryotes, blood cells, and much more. It was not a compound microscope, though. It was more like a very strong magnifying glass. The bacteria that van Leeuwenhoek discovered came from scrapings from an old man's teeth. This man had never brushed his teeth in his life! EW! YUCH!

When and where was he born? *October 24, 1632, in Delft, Holland*

When did he die? *August 30, 1723*

He was inspired after reading a famous book written by Robert Hooke. What is the title? *Micrographia*

It has been speculated that the Dutch painter Johannes Vermeer used optical aids produced by van Leeuwenhoek. How would these have help Vermeer? *It might be that Vermeer used optical aids produced by van Leeuwenhoek to get a better feel for light and perspective in his paintings.*

How many microscopes did van Leeuwenhoek make? What happened to them? *van Leeuwenhoek made over 500 microscopes. Unfortunately, he made them out of silver and gold. When he died, his family sold them for the monetary value of the metal.*

Show What You Know

Types

Multiple Choice

1. A shark is made from *eukaryotic cells*.
2. The bacteria that causes strep throat are made from *prokaryotic cells*.
3. The basic unit of structure and function of an organism is called *a cell*.
4. Unicellular organisms are *prokaryotes and eukaryotes*.

5.

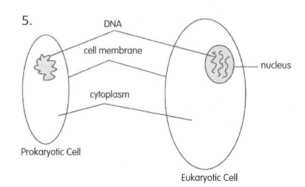

Prokaryotic Cell

Eukaryotic Cell

Fill In the Blanks

The _cell_ theory states:

6. Every _organism_ is made of one or more _cells_.

7. _Cells_ come only from other living _cells_.

8. _Cells_ are the basic unit of _structure_ and _function_ needed to support _life_.

9. **Question:**

What famous scientist coined the term *cell*? Why didn't he see a nucleus, cell membrane, cytoplasm, or genetic material? *Robert Hooke. He didn't see these things because the cells he saw with his microscope were "dead" cells. They were no longer part of a living organism. He saw the cell walls. The nucleus, cell membrane, cytoplasm, and genetic material had decomposed and disintegrated long before he looked at the cork with his microscope.*

10. Match the word with the best definition.

Unicellular ○ —— ○ many-celled

Cell ○ —— ○ a jelly-like material inside all cells

Cell membrane ○ —— ○ genetic material, deoxyribonucleic acid

Cytoplasm ○ —— ○ one-celled

DNA ○ —— ○ an organism whose DNA is located in the cytoplasm

Eukaryote ○ —— ○ the basic unit of life

Multicellular ○ —— ○ an organism whose DNA is located inside the nucleus

Prokaryote ○ —— ○ encloses and protects the inside of the cell

Lesson Review

Cells: Types

Cell theory
- Every organism is made of one or more cells.
- Cells come only from other living cells.
- Cells are the basic unit of structure and function needed to support life.

All cells have
- cell membrane
- cytoplasm
- DNA

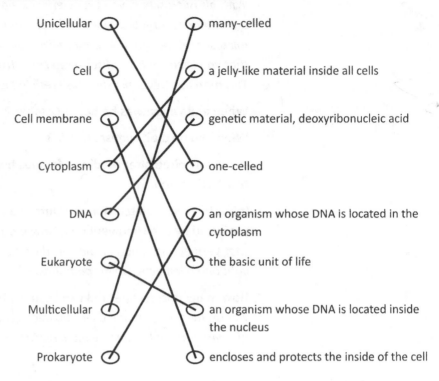

Cells can be prokaryotic = NO nucleus = DNA floats in the cytoplasm
ALL prokaryotes are unicellular. Bacteria are prokaryotic.

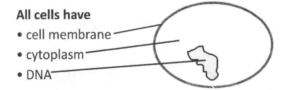

"I'm free!! I'm free!!"

Cells can be eukaryotic = YOU! = DNA in a nucleus
ALL multicellular organisms are eukaryotes. Some unicellular organisms are eukaryotes, too.

"I'm protected!!"

Introduction

How cells work together to make a human is amazing. Even so, it can be dry learning a list of parts and definitions. Memorizing a list of organelles and descriptions of their functions is usually no exception. To make things more interesting and hands-on, students will learn cell anatomy and physiology while conducting what may very well be their favorite lab activity in this course: making a three-dimensional cell out of polymer clay.

Learning Goals

- Learn the names and functions of some of the components of cells, called organelles.

- Explore the specialization of cells in multicellular organisms.

- Describe the similarities and differences between plant and animal cells.

- Learn the technique for making stained slides.

- Learn that cells are 3-dimensional.

- Explore the mathematical explanation for 1-, 2-, and 3-dimensional.

- Research sickle cell anemia.

Extracurricular Resources

Books

Just the Facts, Sickle Cell Disease, Gillie, Oliver

Sickle Cell Disease, Gold, Susan Dudley

Animal Cells: Smallest Units of Life, Stille, Darlene (Prokaryotic cells are smaller than eukaryotic cells so the title is a misnomer, but it's still a good book.)

Plant Cells: The Building Blocks of Plants, Stille, Darlene

The Cell Works: Microexplorers: An Expedition into the Fantastic World of Cells, Bauerle, Patrick and Landa, Norbert

Cell Wars, Balkwill, Frances

Online

Visit Pandia Weblinks for videos and websites recommended for this chapter:

www.pandiapress.com/weblinks-biology2

FSS: Famous Science Series
MSLab: Microscope Lab
SWYK: Show What You Know

Lesson

Cells Specialize

Students are asked to name the organelles in the plant cell during this lesson. Have them study the organelles in the animal cell illustration before doing this activity so they can try to do it from memory. I really like to have students interact with their text in this way. It helps them stay focused on the material and it is a good, quick way to see if they paid enough attention to have retention.

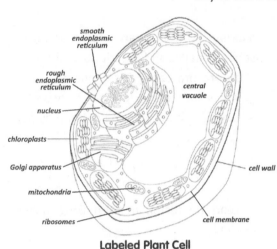

Labeled Plant Cell

This lesson introduces and defines cellular components, called organelles. Similar to how the organs in your body work together so you can function, a cell's organelles work together so a cell can function. Eight organelles common to animal and plant cells, and two organelles found in plant cells but not animal cells, are defined. An explanation of most of the types of organelles and their functions is in the text. I did not include all types of organelles. The specialization of cells in multicellular organisms is explained.

The process most cells use to obtain energy is called cellular respiration. Cellular respiration occurs in the mitochondria. Cellular respiration is covered in Chapter 6.

Chloroplasts absorb energy from the sun. The process is called photosynthesis. Photosynthesis is covered in Chapter 6.

The Golgi apparatus is the only organelle that always starts with an uppercase letter. This is because it is the only organelle named after a person, Camillo Golgi, the scientist who discovered it.

*A note about specialization in multicellular organisms: All the cells that make a multicellular organism have the same DNA in their nucleus. A brain cell, a bone cell, and a red blood cell all have the same DNA; they are all genetically identical. (Until the red blood cell gets rid of its nucleus that is.) What makes a brain cell different from a bone cell and different from a red blood cell are the genes that are turned on in the DNA of each cell type. Different genes code for making different proteins. The different types of proteins (and some other different types of molecules the different cell types make) make different types of cells. So a brain cell has all the same genes, the same DNA, as a bone cell but different sections of DNA, different genes are turned on, producing different types of proteins, and building different types of cells.

Lab

Cells Are Three-Dimensional

Cells are three-dimensional and so are their organelles. That might seem obvious to you. However, a study was conducted to determine if students knew this. Most did not. Every cell they had ever viewed looked flat, so most students thought cells were two-dimensional. To facilitate learning the names, shapes, relative sizes, and three-dimensional structure, students will make their own three-dimensional animal cell this week.

Pandia PRESS

Math This Week

Students measure organelles. Understanding area and 3 dimensions.

3-D Animal Cell

This lab can be a lot of fun. At the end of it, students have a model of a cell that they made themselves. I recommend using Sculpey for the organelles. You can use Play-Doh if the cost of Sculpey is an issue, but Sculpey works really well and lasts longer.

Instructions for making a 3-D model of a plant cell are not included in the student book, but some students have so much fun making the animal cell they make a plant cell too. To make a plant cell model, use the plant cell students labeled in their text for a reference with the following guidelines. For a plant cell:

- Make it square
- The chloroplasts:
  - Are green and organized inside the cell around the outer edge against the cell membrane
  - Are about the same size and shape as mitochondria
- There is a cell wall around the outside of the cell. We made ours out of clay. The plastic container you used to mold the cell would work too.
- There is one central fluid-filled vacuole that is larger than the vacuoles in the animal cell.
- There are no other vacuoles except the central vacuole.
- Include the other organelles that were in the animal cell.

Animal Cell

Plant Cell

Students adding label flags to their animal and plant cells, pushing them into the plaster before it fully hardened.

Answers for Organelle Labels

Nucleus. *This is where the DNA is located. It tells each cell when to reproduce, when to die, and what its job is.*

DNA. *DNA "tells" each cell what its job is and when it is time to divide to make more copies of the cell.*

Smooth Endoplasmic Reticulum. *This is where lipids are made. Cell membranes are made mostly from lipids.*

Mitochondria. *This is where the energy that fuels cells and organisms is made.*

Golgi Apparatus. *This organelle takes proteins from the rough endoplasmic reticulum and assembles them into larger more complex proteins This organelle helps make proteins that are transported and used outside the cell.*

Rough Endoplasmic Reticulum. *This organelle takes proteins made by the ribosomes that are attached to it to the Golgi apparatus. This organelle helps make proteins that are transported and used outside the cell.*

Vacuoles. *A place to store supplies and waste.*

Free-Floating Ribosomes. *Ribosomes build proteins. When they are free-floating in the cytoplasm, the proteins they make are used inside the cell.*

Ribosomes. *Ribosomes build proteins. When they are attached to the rough endoplasmic reticulum, the proteins they make are transported and used outside the cell.*

Cell Membrane. *The organelle that encloses and protects the inside of the cell. It lets material pass into and out of the cell.*

Cytoplasm. *The jelly-like material that is inside of all cells.*

Microscope Lab

Looking at Cells

★ *This lab requires supervision using the knife.*

Yeah! Starting with this week, I am not giving detailed instructions about the procedures for using the microscope. As a result, the procedure portion of the microscope labs becomes much shorter.

This microscope lab has students examining their own cells and those of a plant. Students will look at animal cells, those found inside the student's own cheek, and plant cells from an onion. Students should be able to identify the cytoplasm, nucleus, cell membrane, and for the onion cell, the cell wall.

Students will make a wet mount slide with stain for both specimens. Stain bonds with cellular components, allowing for a much better view of the cell compared to an unstained slide. There are a number of different microscope stains. The type of stain chosen depends on the type of cell and the part of the cell you want to view. Methylene blue is a good stain for looking at the nucleus of cells. This lab is a lot of fun and can be messy. Adult supervision is strongly recommended when cutting with the knife, and when using Methylene blue, stains are indiscriminate.

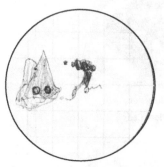

1000x Animal Cell

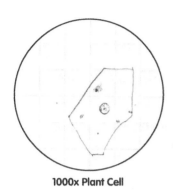

1000x Plant Cell

Famous Science Series

Lab Sheet Answers

Specimen: *cheek cell - animal cell, plant cell - onion*

Type of mount: *wet*

Type of stain used: *methylene blue*

You had to be much more careful working with the animal cells than the plant cells. Why? *Animal cells are more delicate than plant cells because they do not have cell walls.*

Sickle Cell Anemia

Sickle cell anemia is a genetic disorder that changes the shape of the oxygen/carbon dioxide carrying molecule hemoglobin. This changes the shape of the afflicted person's red blood cells from their normal "doughnut shape" to a sickle shape. This disease primarily affects people of West African descent.

Suggested Answers

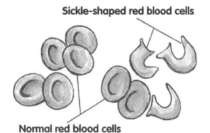

Sickle-shaped red blood cells

Normal red blood cells

How do people get sickle cell anemia? *People with sickle cell anemia inherit it from their parents.*

Why is it a problem when blood cells are not normally shaped? *Normal red blood cells are disk-shaped. There are no pointed parts to clog blood vessels. Sickle- shaped red blood cells are stiff and pointed. The sickle shaped red blood cells can become stuck in narrow blood vessels, causing clogging. This restricts the normal flow of blood and decreases the amount of oxygen getting to the cells in the body. To make things worse, sickle-shaped red blood cells cannot carry as much oxygen in them.*

What health problems do people with sickle cell anemia have? *People with sickle cell anemia may have problems with chest pain, swelling of the hands and feet, an increased rate of infections, fever, or stroke. There are usually painful episodes affecting the bones. Sometimes these episodes require hospitalization.*

Is there a cure for sickle cell anemia? *No. The DNA of a person with sickle cell anemia will keep making misshapen hemoglobin their entire life. The purpose of treatment is pain management and control of the symptoms caused by the disease.*

Show What You Know

The Inside Story

1. Match the organelle with the phrase describing it.

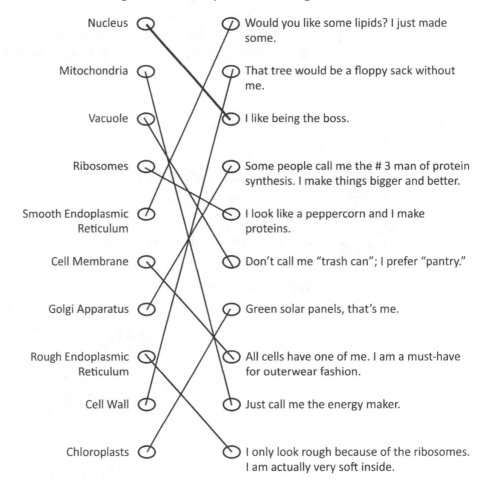

Nucleus

Mitochondria

Vacuole

Ribosomes

Smooth Endoplasmic Reticulum

Cell Membrane

Golgi Apparatus

Rough Endoplasmic Reticulum

Cell Wall

Chloroplasts

Would you like some lipids? I just made some.

That tree would be a floppy sack without me.

I like being the boss.

Some people call me the # 3 man of protein synthesis. I make things bigger and better.

I look like a peppercorn and I make proteins.

Don't call me "trash can"; I prefer "pantry."

Green solar panels, that's me.

All cells have one of me. I am a must-have for outerwear fashion.

Just call me the energy maker.

I only look rough because of the ribosomes. I am actually very soft inside.

2. Below is a list of statements about cell specialization. Write true or false. If false, rewrite the sentence making it true.

F   Unicellular organisms have specialized cells. *Unicellular organisms do not have specialized cells. They have one cell that does it all.*

T   All body cells in an organism have the same DNA, no matter what their specialization.

T   Specialized cells in an organism have different shapes depending on their specialization.

F   All the cells in a multicellular organism have the same number and type of organelles. *The cells in a multicellular organism can have different organelles depending on the type of cell.*

F   One cell from a multicellular organism can survive on its own. *One cell of a multicellular organism cannot survive on its own.*

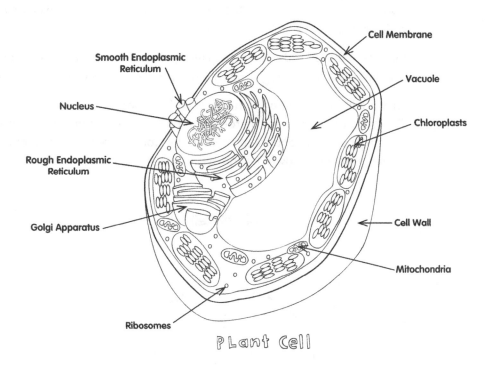

Smooth Endoplasmic
Reticulum

Nucleus

Rough Endoplasmic
Reticulum

Golgi Apparatus

Ribosomes

Cell Membrane

Vacuole

Chloroplasts

Cell Wall

Mitochondria

PLant Cell

Lesson Review

The Inside Story

The cells of multicellular organisms specialize. Different cell types . . .

• look different—different size, different shape
• function differently
• have different amounts of organelles
• have different types of organelles

Cells can have organelles.
Organelles are structures that carry out the functions for a cell.
The organelles and what their purpose is:

  • nucleus—where DNA is in eukaryotes
  • mitochondria—where energy is made, the generator
  • smooth endoplasmic reticulum—where lipids are made; lipids are the
    molecules that make cell membranes
  • vacuole—stores food and waste. In plants, there is a central fluid-filled
    vacuole that helps plants maintain their structure.

Making Proteins

Proteins that are used inside the cell: Ribosomes floating in the cytoplasm make these proteins.

Proteins that are used outside the cell: Three organelles work together to make proteins: ribosomes, rough endoplasmic reticulum, and Golgi apparatus.

They are like conveyer belts in a factory. The ribosomes start making the protein. They send the protein to the rough endoplasmic reticulum, which works on the protein, then they send the protein to the Golgi apparatus, which finishes the work. So the protein is ready to be transported OUT of the cell.

Animal Cell

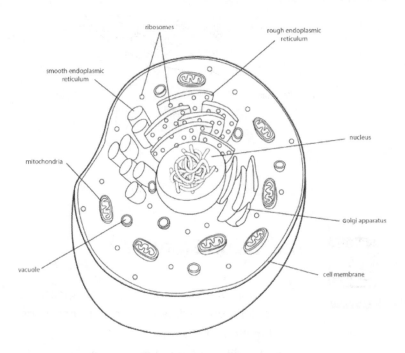

Plant Organelles

Plants have all the organelles animal cells have plus . . .
- chloroplasts–absorb energy from the sun and use it in the process of making food
- cell wall–around the outside of the cell membrane of plant cells. It makes plant cells stiffer than animal cells. It helps gives plants their structure.

Unit II: Cells

Chapter 4: The Chemistry of Biology

WEEKLY SCHEDULE

Two Days

Day 1
- ❑ Lesson
- ❑ Lab
- ❑ Activity
- ❑ FSS

Day 2
- ❑ MSLab
- ❑ Lesson Review
- ❑ SWYK

Three Days

Day 1
- ❑ Lesson
- ❑ Lab

Day 2
- ❑ Activity
- ❑ MSLab

Day 3
- ❑ FSS
- ❑ Lesson Review
- ❑ SWYK

Five Days

Day 1
- ❑ Lesson
- ❑ Lab

Day 2
- ❑ Activity

Day 3
- ❑ MSLab

Day 4
- ❑ FSS

Day 5
- ❑ Lesson Review
- ❑ SWYK

FSS: Famous Science Series
MSLab: Microscope Lab
SWYK: Show What You Know

Introduction

There is a lot of material in this chapter. I recommend using either the three-day or five-day schedule.

Have you ever noticed that the more you learn about different subjects, the more they interrelate? Well, in order to go any further with biology, a little bit of chemistry is essential. Yes, chemistry! The lesson for this chapter starts with a basic introduction to chemistry and how it relates to health and nutrition. Chapter 5 explains movement into and out of cells. Chapter 6 explains photosynthesis. All of these are very important biochemical processes on which all life depends. To teach these concepts on even a basic level, a little bit of chemistry is essential for an understanding of the molecules that make cells and give energy to organisms.

If your student completed RSO Chemistry, this will be a very basic review. If your student has not had chemistry before, this might feel like a lot of information in a short amount of space. If your student has any problems with the chemistry, tutor them a bit, but just keep going. They will understand the chemistry of biology better after applying it to biological concepts as they are presented in subsequent chapters: cell membranes in Chapter 5, photosynthesis in Chapter 6, proteins in Chapter 7, and genetics and DNA in Unit III.

Learning Goals

- Examine how individual atoms link together to make increasingly complex structures that eventually make cells.

- Review/learn some basic chemistry principles.

- Learn the six main elements that make organisms.

- Understand that organisms are made from just a few types of molecules that are cell-building and energy-giving. The raw material for these molecules comes from the food we eat or the liquid we drink.

- Learn about and perform the starch test.

- Learn to read a nutrition label.

- My Food Choices lab identifies six important food groups. Students should recognize which cell-building, energy-making molecules are in which food groups.

- Research about the health problems associated with cholesterol.

Extracurricular Resources

Books

Eat Your Vegetables! Drink Your Milk! Silverstein, Dr. Alvin; Silverstein, Virginia; and Nunn, Laura Silverstein

Physical Fitness, Silverstein, Dr. Alvin; Silverstein, Virginia; and Nunn, Laura Silverstein

Water and Fiber for a Healthy Body, Royston, Angela

Proteins for a Healthy Body, Royston, Angela

Carbohydrates for a Healthy Body, Royston, Angela

Vitamins and Minerals for a Healthy Body, Royston, Angela

Online

Visit Pandia Weblinks for videos and websites recommended for this chapter:

www.pandiapress.com/weblinks-biology2

Lesson

Everything Is Made from Atoms

Often people do not link cell-building and energy-making with the food and drink they ingest. This makes it important to couple the lesson explaining which molecules make cells and energy with information about food. After all, the whole point of a healthy diet is to make sure you get all the molecules your cells need. Along these lines, both labs, the Famous Science Series, and a label reading activity relate the molecules needed by cells to diet. Below is a completed table from the text. When you add them all together (which I encourage you to do) you get 100 percent of the elements that make up a person.

Name of Element	Symbol	Amount in you	Percentage of your body
Hydrogen	H	6,300	*63%*
Oxygen	O	2,550	25.5%
Carbon	C	950	*9.5%*
Nitrogen	N	140	*1.4%*
Calcium	Ca	31	*0.3%*
Phosphorus	P	22	*0.2%*
19 other elements		7	*0.1%*

Lab

My Food Choices

For this lab, students research healthy dietary guidelines online and then create a healthy menu for themselves. Good places to research are the USDA's MyPlate Plan, the Mayo Clinic, and the American Heart Association. The USDA's MyPlate has several advantages over the old Food Pyramid, the main ones being it is a lot easier to understand visually, it separates fruit and vegetables, and starches are no longer the biggest slice (vegetables are). But it does have its critics. For example, there is no place for healthy fats on My Plate.

Your nutrition opinions and practices might be quite different from what is presented on the websites your student researches. I certainly encourage you to discuss these and incorporate them into this lesson. Students are instructed to avoid fad weight-loss diets and those that recommend omitting entire food groups.

Food contains the molecules needed to make cells and get energy. This lab looks at where each type of molecule is within the five food groups. It would be really hard to eat the molecules you need every day for optimum health without some guidance. This lab was written to highlight this.

Activity

Nutrition Labels

Food labels have a lot of good science information on them. They have most of the information needed to make sure you are getting the molecules needed to build cells and make energy.

Worksheet Answers

1. There are 6g of protein in the egg. How many calories in the egg are from protein? *6 grams of protein in an egg x 4 calories in a gram = 24 calories from protein in an egg.*

2. Is whole wheat flour a good source of dietary fiber? *A person who eats about 2,000 calories a day needs 25g of dietary fiber a day. One serving of whole-wheat flour has 4g, or about 16 percent. That is pretty good.*

3. There is a type of fat not listed on the label. What is it? *Unsaturated fat*

4. How many more servings are there in a bag of flour than in a carton of eggs? *76 - 12 = 64 more servings in a bag of flour*

5. How many of the calories in a serving of the whole-wheat flour are from carbohydrates? *21 grams of carbohydrate in a serving of flour x 4 calories in a gram = 84 of 110 calories in a serving of flour are from carbohydrates.*

6. Which has a higher percent Daily Value for protein: flour or eggs? How much higher? *Eggs have a higher percent Daily Value for protein, 6 – 4 = 2 grams higher.*

Microscope Lab

Scrubbing Fiber

In this lab, whole-wheat flour, a good source of dietary fiber, is compared with all-purpose flour, a poor source of dietary fiber. Both whole-wheat flour and all-purpose flour come from wheat. The difference in the two is that the all-purpose flour has had the bran and germ removed. The two flour types are put under the microscope to get a closer look at the difference.

The Starch Test: When iodine comes in contact with starch, a chemical reaction occurs, turning it a blue-black color. Both flour types have starch in them. Both react positively to the starch test. The all-purpose flour is basically all starch. Some parts of the whole-wheat flour are not starch. This shows up clearly when the two different flours are looked at under the microscope.

Suggested Answers

Dietary Fiber.

4g Whole-Wheat	less than 1g fiber All-Purpose

Physical Tests. Note the differences and commonalities:

	Whole-Wheat	All-Purpose
Feel	slightly grainy, rough	soft, silky
Taste	"wheaty" or "nutty"	less taste than whole wheat, sort of slimy
Sight	color varies from brown to whitish tan	white
Smell	no smell or slightly nutty	no smell

Starch Test. Record the color.

	Whole-Wheat	All-Purpose
Flour	brownish tan	white
Iodine	rust color	rust color
Flour & Iodine	blackish blue with bits of brown	blackish blue

Conclusions

Did you confirm the presence of starch in the two types of flour? *yes*

When you looked through the microscope, had all parts of the whole-wheat flour changed color after the starch test? *No, some were still brown.*

What about the all-purpose flour? *All the all-purpose flour turned blackish-blue.*

Whole-wheat flour has the bran and germ; all-purpose flour does not. Both flour types have the endosperm in them. What do the views under the microscope, after the starch test, tell you about the starch content of the bran and germ versus the endosperm? *The endosperm is made from starch. At least part, and maybe all, of the bran and germ are not starch.*

Whole Wheat Grain

Bran

Endosperm

Germ

Famous Science Series

Cholesterol

Your body needs some cholesterol. Too much can be harmful to your health. Students will learn how to identify sources of cholesterol, and methods to prevent high cholesterol.

Suggested Answers

What are the sources of cholesterol? *Animal fat, in particular red meat, dairy products that are not non-fat, and egg yolks*

Is cholesterol found in plants? *No. Food from plants might contain fat, but NOT cholesterol. Foods that come from plants are: grains, oils from plants, beans, nuts, seeds, fruits, and vegetables.*

What are LDL and HDL? Which is bad and which is good? *Your blood uses proteins, called lipoproteins to carry cholesterol. These lipoproteins are called LDL and HDL. LDL are low-density lipoproteins; they are used to measure the bad cholesterol. As LDL (bad) cholesterol circulates, it can slowly build up on the walls of blood vessels that feed the heart and brain. This happens to people with high cholesterol. People with high cholesterol are at a greater risk of having a heart attack or stroke. Not all cholesterol is bad. HDL are high-density lipoproteins; they are good. You want lots of HDL (good) cholesterol and very few LDL (bad) cholesterol in your blood. HDL (good) cholesterol carries cholesterol away from your blood vessels.*

What can you do to avoid having high levels of bad cholesterol?
1. *Exercise*
2. *Eat non-fat dairy products*
3. *Eat high-fiber foods*
4. *When cooking with oil, use olive oil or canola oil.*
5. *Do not eat many foods that are high in cholesterol.*
6. *Do not eat much saturated fat. These raise your levels of LDL.*
7. *Do not eat hydrogenated fats also called trans fats. They raise your levels of LDL. Trans fats are most often found in pre-packaged fried foods and baked goods. Look for the words "hydrogenated oils" or "partially hydrogenated oils" on food labels to determine if trans fats are present.*

Show What You Know

The Chemistry of Biology

In 10 water molecules (10 H_2O), how many hydrogen atoms are there? How many oxygen atoms are there? *# of hydrogen atoms = 2 x 10 = 20 atoms hydrogen, # of oxygen atoms = 1 x 10 = 10 atoms oxygen*

What are the six main elements organisms are made from? *hydrogen, oxygen, carbon, nitrogen, phosphorus, calcium*

Multiple Choice
1. These molecules are the main source of energy for your body: *carbohydrates*
2. Calcium makes bones and teeth strong. Calcium is a *mineral.*
3. Skin, hair, and hemoglobin are all made from these molecules: *proteins*
4. These molecules make DNA: *nucleic acids*
5. These are the molecules that make proteins: *amino acids*
6. These molecules are used to make cell membranes: *lipids*
7. What molecules are needed by your body for chemical reactions to occur in? *water*

8. What are cells are made from? *all of the above*
9. What is a group of all the same atoms is called? *an element*
10. What are the links between atoms called? *bonds*
11. The amount of water in your body is *65%.*
12. What is the serving size of macaroni and cheese? *⅔ cup*
13. If you eat the entire box, how many calories would you consume? *810*
14. What is the second most common ingredient in macaroni and cheese? *cheddar cheese*

Bonus: There is cholesterol in the macaroni and cheese. What ingredient did it come from? Could this product be made cholesterol-free? If yes, explain how. *It comes from the cheese, a dairy product. The product could be made cholesterol free by using cheese made from fat-free milk.*

Lesson Review

The Chemistry of Biology

atoms make molecules → molecules make organelles and cytoplasm → organelles and cytoplasm make cells → cells make organisms

Elements = a group of the same kind of atoms

Organisms are made mainly from 6 elements:

1. Carbon   2. Hydrogen
3. Oxygen   4. Nitrogen
5. Phosphorus 6. Calcium

These are the atoms that make the molecules that make cells that make you

The molecules that make energy and build cells:

- **Carbohydrates** = energy
  - o There are complex carbohydrates and refined carbohydrates.
  - o At least half of all the carbohydrates you eat should be complex carbohydrates.
- **Lipids**
  - o Make cell membranes
  - o Are an energy reserve
  - o Insulate your body and protect your internal organs.
- You have 100,000 different **proteins**.
  - o Proteins have many functions. They make skin, hair, hemoglobin, and muscles, to name a few. Proteins are made from amino acids.
- **Nucleic acids** make DNA.
- Chemical reactions in the body happen in **water**. Chemical reactions are the force driving life.
- **Vitamins and minerals**
  - o Phosphorus is in nucleic acid
  - o Calcium is in bones and teeth
  - o Vitamin C prevents disease like scurvy

Unit II: Cells

Chapter 5: Let's Get Things Moving

WEEKLY SCHEDULE

Two Days

Day 1
- ☐ Lesson
- ☐ Lab day 1
- ☐ MSLab

Day 2
- ☐ Lab day 2
- ☐ Lab report
- ☐ FSS
- ☐ Lesson Review
- ☐ SWYK

Three Days

Day 1
- ☐ Lesson
- ☐ Lab day 1

Day 2
- ☐ Lab day 2
- ☐ Lab report
- ☐ MS Lab

Day 3
- ☐ FSS
- ☐ Lesson Review
- ☐ SWYK

Five Days

Day 1
- ☐ Lesson
- ☐ Lab day 1

Day 2
- ☐ Lab day 2
- ☐ Lab report

Day 3
- ☐ MSLab

Day 4
- ☐ FSS

Day 5
- ☐ Lesson Review
- ☐ SWYK

Introduction

Organisms depend on material getting into and out of cells in order to live. The structure of cell membranes is explained as it relates to cellular transport. The basic transport processes are also defined.

The lab in this chapter is an excellent one for writing a lab report.

Learning Goals

- Learn about the structure of cell membranes and how this structure affects and allows transport across the cell membrane.

- Understand cells use two mechanisms to transport molecules in and out: active and passive transport.

- Recognize that passive transport does not require energy and occurs from areas of high concentration to low concentration.

- Recognize that active transport requires energy and occurs from areas of low concentration to high concentration.

- Learn the types of passive transport: diffusion and osmosis

- Learn two types of active transport: endocytosis and exocytosis

- Research the health risks associated with secondhand smoke.

- Learn more about starch molecules.

- Learn the microscope technique of top lighting.

- Become more proficient at writing lab reports.

Extracurricular Resources

Visit Pandia Weblinks for videos and websites recommended for this chapter:

www.pandiapress.com/weblinks-biology2

Lesson

Move It!

The chemical processes that all organisms rely on for life depend on the transport of molecules and other material into and out of cells. This transport happens across the cell membrane. The cell membrane is a semi-permeable membrane. It is selective about what it lets in and out of the cell.

The text begins by explaining how cell membranes are constructed. The structure of lipid molecules and cell membranes is what makes selective transport across cell membranes possible.

The concepts of diffusion, active transport, and passive transport can be further explained through demonstration. After reading the lesson, I recommend you do so. Here are a few examples:

Diffusion through a gas: Take a bottle of any food extract, or perfume, or other scented liquid (the stronger the scent the better). Open the bottle and let kids smell near the opening. They will smell the liquid, as the scent molecules diffuse out of the bottle. The scent molecules are moving from an area of high concentration, the liquid that is the source of the smell, to an area of low concentration, an area where there were few to none of the scent molecules.

Diffusion through a liquid: Drip two drops of food color into warm water. Watch as the food color goes from an area of high concentration to one of low concentration.

Active and passive transport: Here is a quick way to demonstrate the differences. Have students stand on a secured chair or sofa, something that is not going to fall away when they jump. Have students jump down from the chair. Ask them how much energy it took for them to go from high to low. Now ask students to jump from the floor up onto the chair. (Most students will not be able to do this. Make sure it is safe for them when they jump.) Ask them how much energy it took for them to go from low to high. Discuss the fact that it did not take energy to jump from high to low, so that was passive transport of their body off of the chair. But it took a lot of energy to jump from low to high, so that was an active transport of their body as they attempted to jump onto the chair.

* Protein channels can also be used for active transport. When this is the case, energy is used to transport molecules.

Lab

Diffusion Confusion

This is a diffusion experiment. It takes two or more days to conduct. I recommend students write the lab report, but it is optional. If your student does not write a formal lab report, they could draw a diagram of the experiment and record their observations.

Plastic bags are semi-permeable membranes. There are small holes between the molecules that make the bag. These holes will let small molecules such as water and iodine go through the bag. The holes are not big enough for larger molecules such as the starch molecules in cornstarch to go through. The starch test is used in this experiment to monitor the movement of the molecules.

**Active Transport
In Cells**

Pandia PRESS

Lab Report

Below is an example lab report for the Diffusion Confusion Lab. Your student's hypothesis, observations, results, and conclusions will vary.

Chapter 5: Lab Report

Name: _____   Date: _____

Title: *Diffusion Experiment Across a Plastic Membrane*

Hypothesis

I think the iodine molecules are small enough to go through the plastic membrane of the baggie and the starch molecules are too big to go through it.

Procedure

I mixed two solutions, one with iodine and water and one with cornstarch and water. I poured the cornstarch and water mix into a baggie. A baggie is a semi-permeable membrane. If a molecule is small enough it will diffuse through the baggie.

I put the baggie into a container. I poured the iodine water mix into the container so it surrounded the baggie. If the iodine molecules are small enough to diffuse across the baggie, the cornstarch water solution will get darker. If the cornstarch molecules are small enough to diffuse across the baggie, the iodine solution will get darker.

Observations

10 minutes: the iodine solution was red, the cornstarch solution was white
20 minutes: the iodine solution was red, the cornstarch solution was beginning to gray
30 minutes: the iodine solution was red, the cornstarch solution was slightly greyer
1 hour 30 minutes: the iodine solution was red, the cornstarch solution was a little greyer and red was seeping into the bottom of the baggie and then turning blackish. The cornstarch was settling to the bottom with rust color at the bottom, inside the baggie.
2 hours 30 minutes: same as at 1 hour 30 minutes, except the cornstarch solution was grayer. The next morning: the iodine solution was red. The cornstarch solution was bluish gray. The cornstarch looked rust colored and was settled at the bottom of the baggie.

Results and Calculations

The iodine water solution did not change color. The cornstarch water solution got darker. This shows that iodine molecules are small enough to diffuse into the baggie, but the cornstarch molecules are too big to diffuse out of the baggie.

After the baggie was taken out and cornstarch was mixed into the iodine solution, the solution turned bluish black, indicating a positive result for the starch test. This shows that if cornstarch had been diffusing out of the baggie, the iodine solution would have changed from a rust colored solution to a bluish black one.

Conclusions

I conclude that my hypothesis was correct. Iodine molecules are small enough to diffuse through the semi-permeable plastic membrane of the baggie and cornstarch molecules are too big to go through it.

In Search of Starch

Microscope Lab

★ *Cutting the corn kernel requires adult supervision. Do not let the higher-power lenses touch the specimen.*

The microscope lab for chapter 5 has more relevance as a follow-up to the general lab. For the microscope lab, students use the starch test again. Students take a corn kernel and stain it. When they do this, the starch molecules in the corn become visible. This specimen is less delicate than cheek and onion cells. Therefore, excess stain is much easier to rinse off the specimen. Students experiment with top lighting. The corn slice is thick. This means a light source from above the stage is required. Students make a slice of the specimen with adult supervision.

Possible Answers

1. Did you get a positive result with the starch test? *Yes, part of the kernel turned bluish black after it sat in the iodine.*
2. Did you see cornstarch molecules? If yes, what color were they? *Yes, bluish black*
3. Was the view better with top lighting? *Yes*

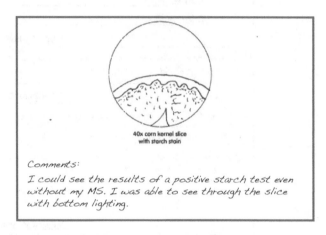

40x corn kernel slice
with starch stain

Comments:
I could see the results of a positive starch test even
without my MS. I was able to see through the slice
with bottom lighting.

Famous Science Series

Secondhand Smoke

When people smoke, the smoke coming from the tip of the cigarette diffuses out as it goes from an area of high concentration to an area of low concentration. This week's Famous Science Series investigates this health risk.

What is secondhand smoking? *Secondhand smoking is also called passive smoking. It is the inhalation of smoke molecules by a nonsmoker. Because of diffusion, even if you are not smoking you will inhale smoke molecules if you are around someone who is smoking. The smoke molecules will diffuse away from the source, over to you, and you will take them into your lungs when you breathe.*

What is the process called that carries the smoke from its source to other areas?

Diffusion. When someone is smoking, the source of the smoke, the object that is being smoked, is surrounded by an area that has a high concentration of smoke molecules. These smoke molecules spread out, diffuse, to areas of low concentration, such as the area around a nonsmoker, a person who isn't creating any smoke molecules. When someone smokes, about half the smoke is inhaled by the smoker; the other half diffuses away from them.

What are the health risks associated with secondhand smoke?

1. *There is an increased risk of cancer compared to someone who has not been exposed to secondhand smoke.*
2. *There is an increased risk of heart disease.*
3. *Secondhand smoke can lead to asthma and other respiratory problems.*
4. *Secondhand smoke increases the risk of sudden infant death syndrome (SIDS).*

**Show What
You Know**

Let's Get Things Moving

1. If you put 5 drops of blue food color in a glass of water, what would happen to the food color? What is this process called? Does it require energy? Draw two pictures of the glass, one right after the food color was dripped into the glass and another an hour later. *The food color would spread throughout the water until it was completely mixed throughout the water. The process is called diffusion. It does not require an input of energy. The pictures should show the food coloring completely diffused into the water after an hour.*

2. In the above scenario, the food color moved from an area of *high* concentration to an area of *low* concentration.

3. Tree roots use passive transport to absorb the water the tree needs. What is this type of passive transport called? *osmosis = the diffusion of water*

4. During active transport, *energy* is needed to move molecules from an area of *low* concentration to an area of *high* concentration.

5. Two examples of active transport are *endocytosis* and *exocytosis*.

6. If your mom cleans your room, do you use your energy to do it? Is it active for you or passive? *no; passive*

7. What if you clean your room? Is this active or passive? *active*

8. *Active transport*–This is the process where cells use energy to move molecules from an area of low concentration to an area of high concentration.
   Diffusion–This is the process where molecules move from an area of high concentration to an area of low concentration.
   Endocytosis–A process that transports large molecules into the cell. Material comes in contact with the cell. Then it is enclosed within its own membrane. Next, it is brought inside the cell and released from the membrane into the cell.
   Exocytosis–A process that transports large molecules out of the cell.
   Material inside the cell is enclosed within its own membrane. It is taken to the cell membrane, with which it then fuses. After that, the material is released out of the cell.
   Hydrophilic–Water-loving
   Hydrophobic–Water-hating
   Lipid bilayer–Made of two layers of lipids.
   Osmosis–The diffusion of water across a membrane.
   Passive transport–The process of moving molecules into and out of cells without using any cellular energy. Diffusion is an example of this.
   Semi-permeable membrane–Lets some but not all things pass through it.

10. Which three of the nine characteristics of life directly relate to the cellular transport? Give a one-sentence explanation for how each works.

1. All organisms take in energy. Food molecules are transported into cells.

2. All organisms get rid of waste. Waste molecules are transported out of cells.

3. All organisms grow. Cells need to transport molecules into them as the raw material for organelles to build cell-growing molecules.

Lesson Review

Let's Get Things Moving

Cell membrane = semi-permeable membrane

The construction of cell membranes allows some molecules to travel across them.

Methods of Cell Transport

1. **Passive transport** = no energy needed

   Diffusion = from high concentration to low concentration

   Osmosis = diffusion of water

   Small molecules diffuse across the lipid bilayer.

   Large molecules diffuse through protein channels.

2. **Active transport** = energy required

   Molecules move from low concentration to high concentration through protein channels.

   Exocytosis = how big molecules exit cells. Inside the cell, the big molecule is engulfed by the cell membrane and released outside the cell.

   Endocytosis = how big molecules get into cells. Outside the cell, the big molecule is engulfed by the cell membrane and released inside the cell.

Unit II: Cells
Chapter 6: Cell Energy

WEEKLY SCHEDULE

Two Days

Day 1
- ❑ Lesson
- ❑ Lab
- ❑ MSLAB

Day 2
- ❑ FSS
- ❑ Lesson Review
- ❑ SWYK
- ❑ Units I & II Exam

Three Days

Day 1
- ❑ Lesson
- ❑ Lab

Day 2
- ❑ MSLab
- ❑ FSS

Day 3
- ❑ Lesson Review
- ❑ SWYK
- ❑ Units I & II Exam

Five Days

Day 1
- ❑ Lesson

Day 2
- ❑ Lab

Day 3
- ❑ MSLab
- ❑ FSS

Day 4
- ❑ Lesson Review
- ❑ SWYK

Day 5
- ❑ Units I & II Exam

FSS: Famous Science Series
MSLab: Microscope Lab
SWYK: Show What You Know

Introduction

This is the end of Unit II on cell biology. There is an exam for this unit plus Unit I in the appendix of the student Workbook. It is optional. You could use the exam simply as a review sheet.

Chapter 5 explained the mechanisms cells use to transport molecules. The subjects of Chapter 6 are photosynthesis and respiration, two important processes that rely on material being transported into and out of cells, and are essential to life as we know it.

Note: Chemical symbols and equations are abstract representations, too abstract for some students. The lab for this week attempts to take out the abstract chemistry of the processes and focus on the concrete biology. That said, I have instructions for writing the chemical equations on the lab sheets. You know your students, though. You might need to tailor this chapter with less chemistry in it, so they understand the process. Pay attention to retention, and work with your students so they understand the processes of photosynthesis and cellular respiration as best as they can.

Learning Goals

- Investigate the scientific method used by van Helmont that began to answer the question of how plants get their cell-building, energy-making molecules.

- Learn to distinguish between heterotrophs and autotrophs.

- Learn the process of photosynthesis, in words and with chemical symbols.

- Learn about chlorophyll.

- Learn the process of cellular respiration, in words and with chemical symbols.

- Recognize that photosynthesis and cellular respiration are a cycle.

- Learn about respiration that occurs when oxygen is not present, called fermentation.

- Learn the types of fermentation: alcohol and lactic acid.

- Research about the photosynthetic organisms that form stromatolites.

- Investigate photosynthesis and cellular respiration in a natural setting.

- View chloroplasts.

Extracurricular Resources

Books

How Did We Find Out About Photosynthesis?, Asimov, Isaac

Top Secret, Gardiner, John Reynolds – This is a fun fiction book about a boy who decides to win the science fair with a project studying human photosynthesis. It is at a fairly easy reading level.

Online

Visit Pandia Weblinks for videos and websites recommended for this chapter:

www.pandiapress.com/weblinks-biology2

Lesson

Math This Week

This lesson introduces chemical equations (stoichiometry).

Let's Eat!

Photosynthesis is the process where plants take the energy from the sun and convert it into stored energy. What is so important about photosynthesis is that the energy from the sun is not consumable energy for life, until photosynthetic organisms convert it into energy held in glucose molecules. All other organisms in one form or another meet their energy needs with this glucose.

There are two types of respiration: cellular respiration and fermentation. Both types of respiration release energy stored in the glucose molecule by breaking down the molecule into smaller molecules.

Photosynthesis and respiration directly relate to three of the characteristics of life, explained in Chapter 1. These are: taking in energy, getting rid of waste, and having some type of respiration.

In this lesson, students study the famous van Helmont experiment that proved that plants do not get their food from soil. Plants make their own food through photosynthesis. This does not mean that soil is not important. Most plants need soil to secure their roots and to have access to water. Soil can also provide mineral nutrients to help make a plant healthier (so it can grow stronger and faster). But soil does not provide them food. This is what van Helmont's experiment proved. You could discuss with your student why a product labeled "plant food" is a misnomer. What would be a more appropriate name?

There are two separate things going on in this chapter regarding the van Helmont experiment. I chose to present the van Helmont experiment in the manner I did as a concrete example of how the scientific method works. This is a very straightforward and simple experiment, but the answers that came out of it are fundamental to all of biology. I hope you will look at and discuss with your students the process and methodology involved separate from the material in the text.

Note about stoichiometry (stoy-key-ahm'-eh-tree): Stoichiometry is the branch of chemistry that deals with the amounts of reactants and products in chemical reactions. Stoichiometry is about balancing chemical equations. In chemical equations, the number of atoms of each element on the product and reactant sides of the equation must match. It is an important aspect of a physical law that says matter cannot be created or destroyed.

For example, in this reaction: $6CO_2 + 6H_2O + energy \longrightarrow C_6H_{12}O_6 + 6O_2$ there are 6 carbon atoms on both sides of the equation, 18 oxygen atoms on both sides of the equation, and 12 hydrogen atoms on both sides of the equation.

Explore

Lab

Energy In, Energy Out

The general lab this week is done outside. If you can, choose a day with some sunlight. The goal of this lab is to help students put the process of the two complementary energy cycles into perspective in the real world and to help them think about how fascinating the whole process is.

This is one of my favorite labs because it brings micro-scale cellular concepts to a larger experiential level. (And it's a great excuse to get outside!) It is very true that photosynthesis and cellular respiration are chemical processes that can be explained by chemical equations that occur on the atomic level. But when it comes to biology, they are so much bigger and more profound than that. For some students, it is important that photosynthesis doesn't get lost in the forest for the trees (pun intended).

Therefore, the ideal place to put the cycle of photosynthesis and cellular respiration in perspective is outside, thinking about what is going on and how it all works. Some students will understand and have no problem with the biochemistry. If you do not have one of those students, sit outside under the sun, feel its energy, and feel the air when you exhale and inhale. Then discuss the process in words. Focus on the big picture, the macro-scale process, without all the chemistry.

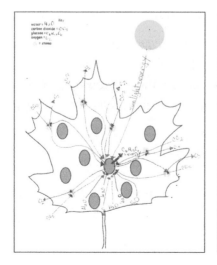
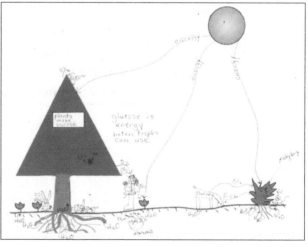

Microscope Lab

Going Green

★ *Cutting the leaf requires adult supervision.*

I experimented with several types of plants. Lettuce and spinach, which are easy to obtain, did not give good results at all. I found that stiff-leaf plants, like orchids and succulents, worked the best because it was easiest to shave a thin slice from a leaf. A thin slice is essential for a clear view.

Microscope hints:

- You will see stomata. Do not confuse the stomata with chloroplasts.

- The chloroplasts through the oil immersion lens were not super clear, but it was fun to look at them.

- If you use the oil immersion lens, make sure you do NOT get oil on the other lenses. Those lenses are not designed to have oil on them. It can ruin them.

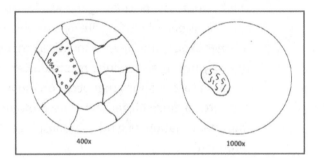

Possible Answers

1. Where are chloroplasts in the cell? Why do you think they are there? *The chloroplasts surround the outer edge of the inner cell. That puts them in the best position to capture sunlight.*

2. Why didn't you use stain to look at the chloroplasts? *The chloroplasts show up well without stain. Plus the stain might make the chloroplasts hard to see.*

3. What is the name of the molecule that makes chloroplasts green? When you look at chloroplasts, do you think you are looking at one or more than one of these molecules? Why? *Chlorophyll. You are looking at many chlorophyll molecules. The molecules are too small to see individually with your light microscope.*

Famous Science Series

Stromatolites

Stromatolites

Stromatolites are formed by photosynthetic bacteria. Plants are not the only photosynthetic organisms. There are also photosynthetic protists and photosynthetic bacteria.

Stromatolite facts:

1. For 2 billion years, cyanobacteria were the dominant form of life on Earth. They used to float in the oceans in huge mats.

2. Stromatolites form when a new layer of cyanobacteria grows on top of an old layer. The older bottom layers die and fossilize.

3. Cyanobacteria are thought to be responsible for the oxygen in the air we breathe. There was plenty of CO_2, sunlight, and water in the atmosphere over 3 billion years ago, but there was not as much oxygen as there is today. The oxygen in the atmosphere comes as a waste product from their photosynthetic activity.

4. Stromatolites can be found in places all over the world. Stromatolite fossils have been found in the bottom layers of the Grand Canyon. A good location to see them is Shark Bay in Western Australia. There you can even see living cyanobacteria on the top layer of stromatolites.

Show What You Know

Cell Energy

1. Not all organisms can photosynthesize. List the group of organisms that can, and state whether they are heterotrophs, autotrophs, or both. *Plants photosynthesize. Plants are autotrophs.*

2. In the boxes below, write the overall chemical reaction for photosynthesis.

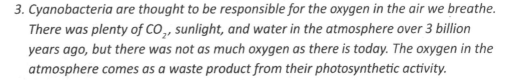

$$6CO_2 + 6H_2O + energy \rightarrow C_6H_{12}O_6 + 6O_2$$

Reactants → Products

3. Name the organelle where photosynthesis happens? *Chloroplasts*

4. Why is this chemical reaction important? *The glucose produced in photosynthesis is the food molecule plants make that harnesses the sun's energy that all organisms use to meet their own energy needs.*

5. What organisms perform cellular respiration? Are they heterotrophs, autotrophs, or both? *All organisms use cellular respiration; therefore they are both heterotrophs and autotrophs.*

6. In the boxes below, write the overall chemical reaction for cellular respiration.

$$6O_2 + C_6H_{12}O_6 \rightarrow 6H_2O + 6CO_2 + energy$$

Reactants → Products

7. Name the organelle where cellular respiration happens. *Mitochondria*

8. Why is this chemical reaction important? *This is the chemical reaction that organisms use to get the energy needed to sustain life.*

9. Photosynthesis and cellular respiration are a called a cycle. Why? *The reactants used in photosynthesis are 6 carbon dioxide molecules, 6 water molecules, and energy from the sun; the products of photosynthesis are 1 glucose molecule and 6 oxygen molecules. The reactants in cellular respiration are 1 glucose molecule and 6 oxygen molecules; the products are 6 carbon dioxide molecules, 6 water molecules, and energy that the organism can use. The only thing that doesn't cycle is the energy.*

10. Match the word with the correct definition.

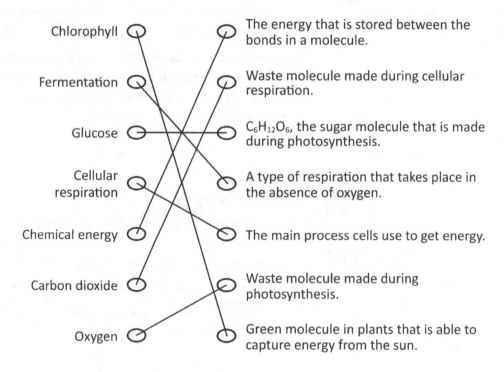

11. Compare and contrast lactic acid fermentation with alcoholic fermentation.
    Compare: Both 1. are types of fermentation, 2. occur when no oxygen is present, 3. occur in the cytoplasm 4. generate less energy than cellular respiration.
    Contrast: Lactic acid fermentation produces lactic acid. Yogurt and cheese are made using lactic acid fermentation. If your muscles run low on oxygen when you are exercising, they use this type of fermentation. Alcoholic fermentation produces alcohol and carbon dioxide. This type of fermentation results in yeast bread rising.

12. What is the significance of van Helmont's experiment? *Van Helmont's experiment proved that plants do not get the food they need to grow from the soil.*

Lesson Review

Cell Energy

All organisms . . .

1. Take in energy, starting with photosynthesis.

2. Get rid of waste. In photosynthesis it is oxygen.

In cellular respiration, it is carbon dioxide.

3. Have some type of respiration, cellular respiration and fermentation.

Chloroplasts = organelle where photosynthesis happens

Chlorophyll = the green molecules in chloroplasts that absorb energy from the sun

Photosynthesis = The energy from the sun is converted into energy that all organisms can use.

carbon dioxide + water + sunlight $\rightarrow$ glucose + oxygen

$6CO_2 + 6H_2O$ + energy $\rightarrow$ $C_6H_{12}O_6 + 6O_2$

Glucose

- Plants take energy from the sun and store it in the bonds that hold glucose together.
- Glucose is a type of sugar.
- Sugar is a type of carbohydrate.
- When plants do not use all the glucose they make, they turn glucose into starch.

Cellular Respiration

- How organisms get the energy stored during photosynthesis.
- Starts in the cytoplasm, finishes in mitochondria.
- Must have oxygen.
- Cells get energy from glucose by breaking the bonds holding the molecule together.

oxygen + glucose $\rightarrow$ water + carbon dioxide + energy

$6O_2 + C_6H_{12}O_6 \rightarrow 6H_2O + 6CO_2$ + energy

Photosynthesis and cellular respiration are a cycle!

Fermentation

- Respiration when there is no oxygen
- In the cytoplasm
- Less energy than cellular respiration
- There are two types: 1. alcohol fermentation 2. lactic acid fermentation

Teacher Guide

Units I and II: Organisms and Cells
Answer Key Unit Exam Chapters 1–6

The exam for Units I and II is found in the appendix of the student textbook.

Multiple Choice (2.5 points each, 30 points total)

The organelle where photosynthesis occurs is the <u>chloroplast</u>.

An organism whose DNA is in a nucleus is a <u>eukaryote</u>.

The six main elements organisms are made from are <u>hydrogen, oxygen, carbon, nitrogen, phosphorus, calcium</u>.

The building block of all living things is the <u>cell</u>.

Semi-permeable membranes <u>let some but not all things pass through</u>.

This organelle gives plants the structure they need to stand tall. <u>Cell wall</u>

Because the cells of multicellular organisms specialize, <u>there are many different sizes and shapes</u>.

Fermentation– <u>all of the above</u>

Passive transport <u>does not use energy</u>.

The cell theory states that– <u>all of the above</u>

Some scientists think viruses should be reclassified as organisms because they <u>are able to reproduce</u>.

After a virus has infected a cell, it turns the cell into <u>a virus-making factory</u>.

2. **Vocabulary**

Match the word with the definition that best fits. (1.5 points each, 18 total)

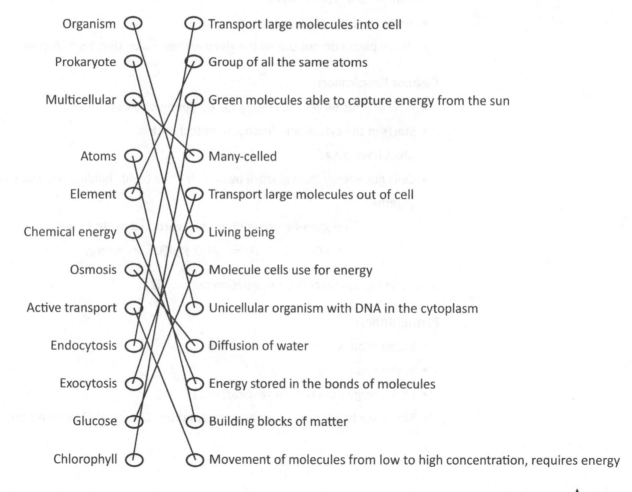

Organism	Transport large molecules into cell
Prokaryote	Group of all the same atoms
Multicellular	Green molecules able to capture energy from the sun
Atoms	Many-celled
Element	Transport large molecules out of cell
Chemical energy	Living being
Osmosis	Molecule cells use for energy
Active transport	Unicellular organism with DNA in the cytoplasm
Endocytosis	Diffusion of water
Exocytosis	Energy stored in the bonds of molecules
Glucose	Building blocks of matter
Chlorophyll	Movement of molecules from low to high concentration, requires energy

Pandia PRESS

3. **Fill in the blanks.** (2 points each, 8 points total)

Proteins are made from <u>amino acid</u> molecules.

Organisms eat <u>carbohydrate</u> molecules for energy. (glucose would also be correct)

Chemical reactions in an organism happen in <u>water</u>.

<u>Protein</u> molecules make hair, skin, and hemoglobin.

4. **Draw a cell membrane.** Include a protein with a protein channel. Label all the parts. (5 points) Show diffusion across the lipid bilayer. (4 points)

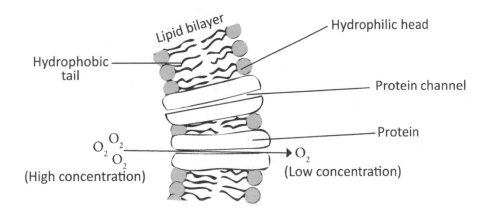

5. **Write the chemical reaction for photosynthesis and cellular respiration.** Names of the molecules can be in words or chemical symbols. Give the number of each type of molecule. (1.5 points each for correct answer, 15 points total)

<u>Photosynthesis</u>

$$6CO_2 + 6H_2O + \text{energy} \rightarrow C_6H_{12}O_6 + 6O_2$$

6 carbon dioxide + 6 water + sunlight → 1 glucose + 6 oxygen

<u>Cellular respiration</u>

$$6O_2 + C_6H_{12}O_6 \rightarrow 6H_2O + 6CO_2 + \text{energy}$$

6 oxygen + 1 glucose → 6 water + 6 carbon dioxide + energy

6. **Organelles**. Fill in the names of the organelles and other material in the cell below. Use their location and description as a guide. (2 points each, 20 points total)

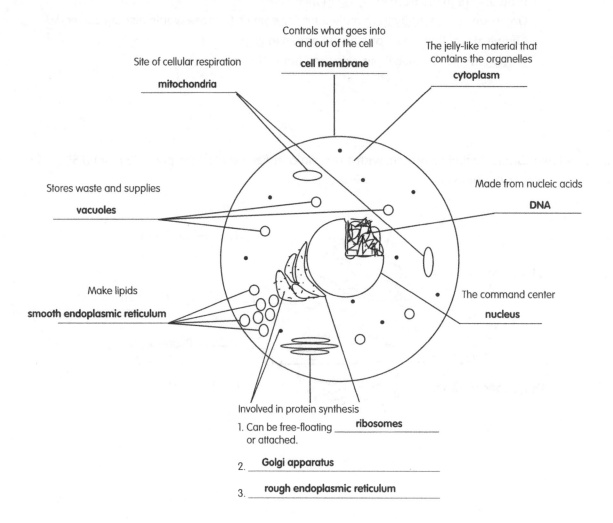

What two organelles does a plant cell have that an animal doesn't? A cell wall and chloroplasts

7. **Extra Credit. There are nine characteristics all organisms have in common. What are they?**
(1 point for each characteristic, 9 extra credit points total)

1. All organisms take in energy
2. All organisms get rid of waste.
3. All organisms move.
4. All organisms grow.
5. All organisms reproduce.
6. All organisms respond to their environment.
7. All organisms have some type of circulation.
8. All organisms have some type of respiration.
9. All organisms are made of one or more cells.

(/100) x 100 =   +   extra credit points =

 Pandia PRESS

WEEKLY SCHEDULE

Two Days

Day 1
- ❑ Lesson & Activity
- ❑ Poetry
- ❑ Lab

Day 2
- ❑ MSLab
- ❑ FSS
- ❑ Lesson Review
- ❑ SWYK

Three Days

Day 1
- ❑ Lesson & Activity
- ❑ Poetry
- ❑ Lab

Day 2
- ❑ MSLab
- ❑ FSS

Day 3
- ❑ Lesson Review
- ❑ SWYK

Five Days

Day 1
- ❑ Lesson & Activity
- ❑ Poetry

Day 2
- ❑ Lab

Day 3
- ❑ MSLab

Day 4
- ❑ FSS

Day 5
- ❑ Lesson Review
- ❑ SWYK

FSS: Famous Science Series
MSLab: Microscope Lab
SWYK: Show What You Know

Introduction Unit III

Every cell at some time in its life has DNA. The genetics unit explains the structure and function of DNA and how cells maintain the correct amount of DNA. It also describes how and why genetic variability occurs.

Chapter 7 describes the structure and function of DNA.

Chapter 8 explains the cell cycle for somatic (body) cells, the process of cell division for these cells, and how cells make more somatic cells.

Chapter 9 details the process (used by organisms that reproduce sexually) of dividing DNA in half to create male and female gametes.

Chapter 10 defines the term *trait* and then examines how the variability of traits occurs. This chapter also explains how to use a Punnett square, a table that predicts the likely outcome of a trait based on the genetic makeup (genotype) of an organism.

Introduction Chapter 7

Chapter 7 describes the structure and primary function of DNA in eukaryotic cells. The approach I use to teach the structure is to go step by step, starting with the nucleotide bases that make up the DNA molecule and ending with a completed molecule of DNA, the chromosome. The last part of chapter 7 explains how DNA, which cannot leave the nucleus, transfers the information it holds to RNA. This holding and transfer of information is the primary function of DNA.

Learning Goals

- Understand the basic shape and structure of a molecule of DNA.

- Recognize the relationship between chromosomes and genes.

- Learn the basic process used to build proteins.

- Learn how DNA replicates.

- Research the two primary scientists who determined the structure of DNA.

- Learn the purpose and function of RNA.

Extracurricular Resources

Books

Have a Nice DNA, Balkwill, Frances R., Ralph, Mic

Gene Machines, Balkwill, Frances R., Ralph, Mic

DNA Analysis: Forensic Fluids & Follicles, Hamilton, Sue

Understanding DNA: A Breakthrough in Medicine, Allan, Tony

Francis Crick and James Watson: Pioneers in DNA Research, Bankston, John

John Bankston (Author) - many books written by this author on the subject of DNA

Francis Crick and James Watson: And The Building Blocks of Life, Edelson, Edward

Online

Visit Pandia Weblinks for videos and websites recommended for this chapter:

www.pandiapress.com/weblinks-biology2

Lesson

What in the World Is DNA?

This lesson takes students step by step through an explanation of DNA. Students interact with the material by coloring the components of DNA in their Workbook. At the point where I switch from discussing structure to discussing function, I emphasize that chromosomes are made from genes and that each gene codes for making one protein. In reality, some proteins, such as hemoglobin, are made from multiple genes, each coding for part of the protein. I also explain that when the chromosome uncoils to make a protein it only uncoils one gene in length as opposed to when it uncoils during nuclear division, the topic of chapters 8 and 9. During nuclear division, the chromosome uncoils along its entire length as a copy is made along each side and a duplicate chromosome is made.

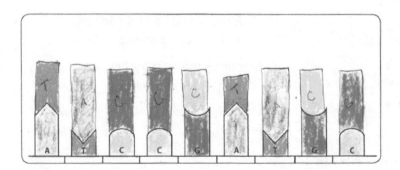

Poetry

The Fickle Base

"The Fickle Base" is a poem that highlights the differences and similarities in base pairing between RNA and DNA. It also emphasizes the basic structure of DNA. The poem can be read as part of the lesson, transcribed in your student's notebook, or perhaps recited aloud as a separate activity. It might also be helpful to use the poem as review throughout teaching this unit. Use this poem as it best works for your student.

Lab

Marshmallow DNA

Students make a marshmallow model of DNA that is open so a strand of RNA can be transcribed along one side. If you are careful with the model and lay it flat, it will dry out and last a long time.

I like using marshmallows to make models because they are cheap, easy to get in multiple colors, and very pliable. This model could also be made with beads or Styrofoam balls.

Microscope Lab

Looking at My DNA

This lab requires some preparation the night before.

The long strand of DNA is a gray white blob even with the microscope. It is more the coolness factor of seeing your own DNA that is so neat.

There are several steps involved in getting DNA out of the cell and the nuclear membrane and then separating it from the proteins it is wrapped around. This is the most involved chemical process of any experiment so far. Students need to follow each step carefully.

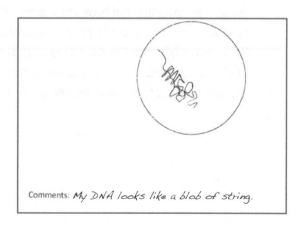

Comments: *My DNA looks like a blob of string.*

Famous Science Series

Watson and Crick

The story of how the structure of DNA was elucidated is fascinating. There are several adult nonfiction books devoted to the subject. I do not go into any of that, but if you and your student are interested I suggest you conduct an Internet search. The focus in this course is on two of the three scientists, James Watson and Francis Crick, who won the Nobel Prize for determining the structure of DNA. They were helped by Maurice Wilkins and Rosalind Franklin, but Watson and Crick were the principal scientists who put the information together.

James D. Watson, Francis Crick, and Maurice Wilkins won the Nobel Prize in 1962. What did they do to earn this award? *They determined the structure of the DNA molecule. By studying the structure, Watson and Crick were able to figure out how DNA molecules could transfer genetic information. Using an X-ray diffraction photograph taken by Rosalind Franklin, Watson and Crick determined that DNA had a double helix structure.*

Why wasn't Rosalind Franklin part of the group who won the Nobel Prize? Do you think that is fair? *The Nobel Prize is only awarded to living persons. Rosalind Franklin had died in 1958. Opinions will vary to the second question.*

When and where did Watson and Crick meet? *They met at Cambridge in 1951.*

<u>James D. Watson</u>

When and where was he born? *April 6, 1928, in Chicago, Illinois*

He had a special kind of memory. What was it? *Watson had a photographic memory.*

He was a regular contestant on what radio show? When was his first appearance? *By the age of ten, he was a regular contestant on the radio show Quiz Kids.*

How old was he when he started college? What college did he attend? *He was 15 when he began attending the University of Chicago.*

<u>Francis Crick</u>

When and where was he born? *June 8, 1916, in Northampton, England*

How old was he when he told his mother he wanted to be a scientist? *12*

What was The Blitz and how did it affect Crick's work? *During WWII, in the summer of 1940, the German air force was regularly flying over London and dropping bombs on it; this was called The Blitz. One of those bombs dropped through the roof of Crick's lab and destroyed his experimental apparatus. After that, he went to work designing mines to use against the Germans.*

Show What You Know

Double Helix Key

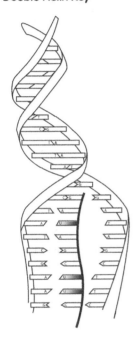

The Message

1. Base pairs in DNA connect to each other with *hydrogen bonds*.
2. Bases in DNA are held together with a backbone of *sugar phosphate chains*.
3. A chromosome is made of *one molecule of DNA*.
4. DNA is shaped like *a double helix*.
5. One gene has instructions for making *one protein*.
6. One RNA molecule codes for the synthesis of *one protein*.
7. One codon has the instructions for making *one amino acid*.
8. The DNA in your cells *is different than your parents*.
9. To make a protein *one gene that codes for making the protein unravels*.
10. The transcription of a molecule of RNA is most like the process of *replication*.
11. *ribosomes, rough endoplasmic reticulum, Golgi apparatus*
12. One hemoglobin molecule has 574 amino acids in it. How many base pairs does it take to make the amino acids in a hemoglobin molecule? How many codons?

    574 x 3 = 1,722 base pairs make hemoglobin

    574 codons make hemoglobin
13. The opposite strand of DNA is called the complementary strand. Make the complementary strand of DNA below. (Refer to the Double Helix Key.)

    A C G T T A G C C G A T

    T G C A A T C G G C T A
14. Make a complementary strand of RNA. (Refer to the Double Helix Key.)

    A C G T T A G C C G A T

    U G C A A U C G G C U A
15. Put these in order as you build a chromosome.

    gene, nucleotide base, codon, base pair → chromosome

    nucleotide base → base pair → codon → gene → chromosome
16. Replication: *making a copy of DNA along a complementary strand of DNA*

    Transcription: *building an RNA molecule along a gene sequence of DNA*

    Translation: *transfer of information from an RNA molecule to a ribosome, to make a protein*
17. List four reasons RNA is used for protein synthesis.

    RNA is: 1. one gene long

    2. single stranded

    3. fits through nuclear membrane (which DNA cannot)

    4. can be built along a strand of DNA

Bonus Questions:

Do the ribosomes in humans make all 100,000 types of proteins your body makes? Support your answer. *There are eight essential amino acids your ribosomes do not make. You must get these eight amino acids from the foods you eat.*

Plants also use 20 different amino acids to build proteins. How do plants get all the amino acids they need? *Since plants do not eat, they must make all the amino acids they need to make proteins.*

Lesson Review

The Message

★ All cells have DNA in them, except mature red blood cells.
★ DNA is made from nucleotide bases held together by a chain of sugar and phosphate molecules.

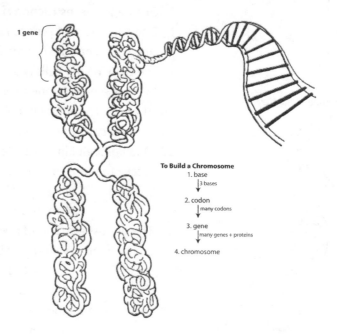

1 gene

To Build a Chromosome
1. base
   | 3 bases
2. codon
   | many codons
3. gene
   | many genes + proteins
4. chromosome

One molecule of DNA: • has two strands with the bases bonded together by hydrogen

• is shaped like a double helix

In DNA, adenine pairs with thymine, and cytosine pairs with guanine.

DNA can replicate = uncoil and build a complementary strand of DNA with the uncoiled strand as a template.

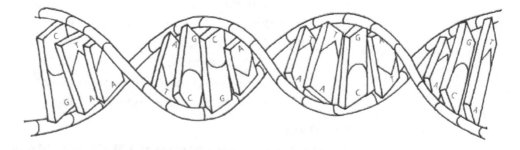

The function of DNA = holding and passing along the code for making proteins
   one DNA molecule = one chromosome
   one chromosome = many genes
   one gene makes one protein

Pandia PRESS

Review from previous chapters: Proteins are made from amino acids. Protein synthesis starts with ribosomes (organelles that are in the cytoplasm outside the nucleus).

RNA solves the problem of DNA being too big to leave the nucleus.

The differences between RNA and DNA:
1. RNA is single-stranded.
2. RNA is one gene long, much shorter than DNA.
3. RNA fits through the membrane around the nucleus.
4. RNA uses another base, uracil, instead of thymine to bond with adenine.

The process:
1. A segment of DNA, which is one gene long, uncoils.
2. Transcription = A molecule of RNA is built along the segment of DNA.
3. RNA leaves the nucleus and attaches to a ribosome.
4. Translation = The code from the RNA is translated to a ribosome and is used to build a protein.

Review from previous chapters: If the protein is being made for inside the cell, it is made by the ribosome alone. If the protein is being made for outside the cell, it goes from the ribosome → rough endoplasmic reticulum → Golgi apparatus.

Unit III: Genetics

Chapter 8: Mitosis–One Makes Two

WEEKLY SCHEDULE

Two Days

Day 1
- ❏ Lesson
- ❏ Lab
- ❏ FSS

Day 2
- ❏ MSLab
- ❏ Lesson Review
- ❏ SWYK

Three Days

Day 1
- ❏ Lesson
- ❏ Lab

Day 2
- ❏ MSLab
- ❏ FSS

Day 3
- ❏ Lesson Review
- ❏ SWYK

Five Days

Day 1
- ❏ Lesson

Day 2
- ❏ Lab

Day 3
- ❏ MSLab

Day 4
- ❏ FSS

Day 5
- ❏ Lesson Review
- ❏ SWYK

Introduction

Chapters 8 and 9 describe the processes cells use to maintain the correct type and number of chromosomes. All the functions of an organism happen at the cellular level; that makes it important for cells to have just the right information they need—not more, not less, and not different.

Learning Goals

- Learn the three steps of the life cycle of a somatic cell.

- Learn about the three phases of interphase.

- Recognize the need for cells to make copies for growth, repair, and replacement.

- Investigate further the shape of chromosomes, and how and why chromosomes replicate.

- Learn what happens during the four steps of mitosis.

- Understand that the process of mitosis leads to genetically identical cells.

- Learn about the process of cytokinesis.

- Understand the differences and similarities of cytokinesis in animal and plant cells.

- Learn that humans have 46 chromosomes in their somatic cells and where they come from.

- Familiarize yourself with the terms ploidy, haploid, and diploid.

- Learn how some organisms use mitosis and cytokinesis to reproduce asexually.

- Research stem cells.

- Investigate and view cells undergoing mitosis.

Extracurricular Resources

Visit Pandia Weblinks for videos and websites recommended for this chapter:

www.pandiapress.com/weblinks-biology2

FSS: Famous Science Series
MSLab: Microscope Lab
SWYK: Show What You Know

Lesson

Carbon Copies

- This chapter has a lot of unfamiliar vocabulary. Pay attention to the new words as you go along, so that the vocabulary does not keep students from understanding the concepts.
- Mitosis is a visual process. Use the acronym PMAT, pronounced "P mat." Visualize each step with one of the letters.
- The terms "mitosis" and "cell division" are often used interchangeably when referring to the cell division of somatic cells, sometimes even in science texts. These two terms are NOT synonyms. Mitosis is the nuclear division that occurs in somatic cells, which is part of cell division in somatic cells but not the entire process.
- The DNA in an organism makes it who and what it is. Changes to DNA, mutations, drive evolution. Mutations are rarely beneficial to individual organisms, though. Mutations are discussed in detail in Unit V, the Evolution Unit.
- Some bacteria reproduce sexually, with an exchange of DNA. I do not discuss that in this course.

Math This Week

Math is infused through all parts of this and the next chapter. The subject of both chapters is cell division. Cells must make copies. When they do, the chromosome number must stay constant and, in the case of somatic cells, genetically identical. The processes of replication and mitosis perform that function.

Ploidy, haploid, and diploid are all math terms referring to the number of sets of chromosomes in an organism's cells. Most organisms are VERY sensitive to the number of chromosomes. Some groups of plants are less sensitive, however, such as seedless watermelons. Seedless watermelons have been genetically modified so that they have three copies of chromosomes, making them triploids.

Lab

Life Cycle of a Cell Poster

This week students make their own poster showing the cell cycle. It can be very beneficial for students to make their own reference material. It gives them ownership of the material, plus they have it to help them remember how the cell cycle works. I have included photos of what the two posters I made look like. They are just for reference.

Example of Poster 1

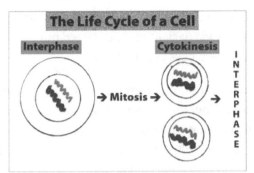

Example of Poster 2

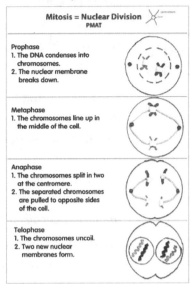

Microscope Lab

Mitosis with a Microscope

When I planned this chapter, it was with the thought that students would make their own slide of an onion tip. I spent quite a bit of time trying to get a good slice. I never saw even one nucleus undergoing mitosis. So, I purchased a prepared slide showing the phases of mitosis. It was worth it. The slide shows many examples of nuclei going through all the phases of mitosis.

Viewing cells under a microscope takes practice and patience. The cells are not going to look exactly like the drawings of each stage; nature isn't so neat and tidy. But if you study the slides long enough, you'll begin to pick up the differences between the cells and differentiate the stages.

Below are examples of the stages of mitosis as seen under the microscope.

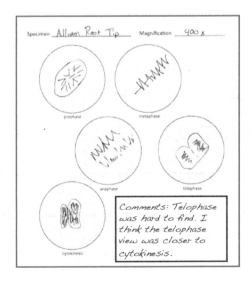

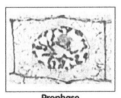
Prophase

Metaphase

Anaphase

Telophase

**Famous Science
Series**

Stem Cells

If you have a particularly inquisitive student, he might have already wondered how cells in his body can be so different and specialized if he started out as one cell that produced identical copies of itself. Stem cells are the answer.

The first cell that becomes a multicellular organism is an embryonic stem cell. It divides, becoming more stem cells. How do a small amount of embryonic stem cells become all the specialized cells that make a human? *In a 3- to 5-day old human embryo, approximately 30 stem cells, called embryonic stem cells, begin to differentiate into the specialized cells of the developing fetus. These 30 cells divide to become all the cell types that become the organs, like the heart and lungs. The cells are all genetically identical, but they begin to specialize.*

What is the purpose of adult stem cells? *Adult stem cells are found in all humans including children. Adult stem cells repair and replace cells lost through use, injury, and disease. Some examples of cells generated from stem cells are those found in bone marrow, blood, muscles, skin, and the brain.*

Stem cells are being used to treat cancer of the blood cells, called leukemia. How? *Doctors inject stem cells from a healthy person's bone marrow into the patient's bloodstream. If successful, the stem cells make new, healthy cells in the patient's bone marrow that replace the cancer-ridden cells.*

What other diseases do researchers hope stem cells can help cure? *Regenerative medicine is a new field of science that studies stem cells and their potential to treat and cure diseases. Because stem cells can become other specialized cell types, there is the potential that they can be used to treat diabetes, Alzheimer's disease, Parkinson's disease, blindness, deafness, cancer, cardiovascular disease, and spinal cord injuries. There are many more people in need of organ transplants than there are organs available. One day, researchers might be able to make those organs in a lab starting with stem cells and directing them to become the specialized cells of that organ's tissue.*

**Show What
You Know**

Mitosis—One Makes Two

1. Chromosomes are held together by *centromeres*.
2. Most of the cell cycle is spent in this phase: *interphase*
3. The phases of mitosis go in the following order: *prophase, metaphase, anaphase, telophase*
4. When DNA copies itself to make two genetically identical chromatids, it is called *replication*.
5. The two genetically identical copies of a chromosome that are connected by a centromere are called *chromatids*.
6. Most bacteria reproduce asexually using the following process: *binary fission*

7. What are the three main stages of the cell cycle: *interphase, mitosis, cytokinesis*

8. Interphase is divided into three parts. What happens in each part?
    G1 Phase: *The cell does its job; it also grows bigger and duplicates its organelles.*
    S Phase: *Inside the nucleus, copies of DNA are made.*
    G2 Phase: *The cells grows more and prepares for mitosis.*

9. Humans are (haploid, *diploid*) organisms. This means their ploidy is (1, *2*).
    Humans have (23, *46*) chromosomes in their somatic cells. The chromosomes
    are (single, *in pairs*). Humans get (all, *half*) their chromosomes from their
    mother.

10. A somatic cell with 8 chromosomes divides. How many daughter cells are
    produced and how many chromosomes are in each daughter cell? *2 daughter
    cells, with 8 chromosomes in each*

11. The daughter cells are *genetically* identical to each other with the same number
    of *chromosomes* in their cells.

12. What are the phases of mitosis this cell is in? Write the answer next to the cell.
    Number the phases according to the correct order they occur.

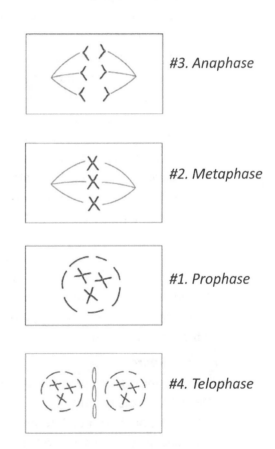

#3. Anaphase

#2. Metaphase

#1. Prophase

#4. Telophase

What type of eukaryotic cell is this, plant or animal? How do you know? *It must be a
plant cell because a cell plate is forming during telophase.*

Lesson Review

Mitosis–One Makes Two

Cell theory

1. Cells come from other living cells.
2. Cells perform all the functions that makes an organism alive.

Somatic = body cells. Almost all your cells are somatic.

Somatic cells divide so an organism can grow, heal wounds, and replace cells that have died. When somatic cells divide they need to keep the same type and number of chromosomes.

Life cycle of somatic cells

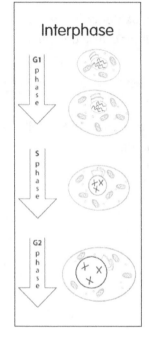

Interphase

- ★ interphase
- ★ mitosis
- ★ cytokinesis
- ★ Interphase
  - • G1 phase = cell does its job
  - • S phase = DNA is replicated
  - • G2 phase = cell prepares for mitosis
- ★ Mitosis = PMAT
  - • Prophase–chromosomes condense and nuclear membrane breaks down
  - • Metaphase–chromosomes line up down the middle
  - • Anaphase–chromosomes split at centromere & separated chromosomes are pulled to opposite sides
  - • Telophase–chromosomes uncoil & two nuclear membranes form, one around each set
- ★ Cytokinesis–the cell divides into two cells
  - • Plant cells must form a new cell wall and new cell membrane. These cells are GENETICALLY IDENTICAL!

Ploidy = number of sets of chromosomes

haploid = 1 set

diploid = 2 sets

Humans are diploid. Humans have 2 sets of chromosomes. Humans have 46 chromosomes: 23 from mother, 23 from father = 2 sets. Humans do NOT use mitosis for reproduction.

Asexual Reproduction

Some organisms do reproduce with mitosis. The offspring are GENETICALLY IDENTICAL to the parents.

- • Binary fission (bacteria)–interphase followed by cytokinesis, 2 new daughter cells
- • Budding–a bud forms, falls off, and becomes a new organism
- • Vegetative propagation–a new plant forms from the growth
- • Regeneration–a new organism can grow or repair a lost body part

Teacher Guide

Unit III: Genetics

Chapter 9: Meiosis Divides by Two and Makes You

WEEKLY SCHEDULE

Two Days

Day 1
- ❏ Lesson
- ❏ Lab
- ❏ FSS

Day 2
- ❏ MSLab
- ❏ Lesson Review
- ❏ SWYK

Three Days

Day 1
- ❏ Lesson
- ❏ Lab

Day 2
- ❏ MSLab
- ❏ FSS

Day 3
- ❏ Lesson Review
- ❏ SWYK

Five Days

Day 1
- ❏ Lesson

Day 2
- ❏ Lab

Day 3
- ❏ MSLab

Day 4
- ❏ FSS

Day 5
- ❏ Lesson Review
- ❏ SWYK

Introduction

Chapter 9 is about meiosis. Meiosis is the process of cell division that results in cells that contain half the number of chromosomes. It is a necessary process for sexual reproduction. I have chosen to present meiosis as a personal story to your student or child. He will read about meiosis as I describe the process of cell division that happened in his parents before he was conceived, and the subsequent fertilization which created the zygote that became him. This chapter (combined with reproductive anatomy taught in Chapter 16) could provide an excellent introduction and opportunity for sex education.

Learning Goals

- Recognize that sexually reproducing organisms need a mechanism to halve the number of chromosomes in the cells they use to create new organisms.

- Further familiarity with the genetics terminology and concept of ploidy and n.

- Remember that humans are diploid and have 2 sets of 23 chromosomes.

- Learn that sexually reproducing organisms have haploid sex cells, called gametes: sperm and eggs.

- Define zygote.

- Learn the steps and basic processes occurring at each step for meiosis.

- Visualize and understand how two haploid cells, one from the mother and one from the father, fuse during fertilization to form a diploid organism.

- Learn that gender is determined by the father's gamete.

- Learn that people who are XX at the 23rd chromosome are female, and those who are XY are male.

- Recognize the differences between asexual and sexual reproduction.

- Research the genetic disorder Down syndrome.

Extracurricular Resources

Visit Pandia Weblinks for videos and websites recommended for this chapter:

www.pandiapress.com/weblinks-biology2

Lesson

Math This Week

More multiplying and dividing by 2. Sex cells start as 1 cell and end as 4 cells.

Meiosis Makes You Unique

Instructor Notes: In humans: <u>Mitosis</u> results in 2 genetically identical cells, both with 46 chromosomes. <u>Meiosis</u> results in 4 genetically variable cells, each with 23 chromosomes. That is one half the number of chromosomes in the nucleus of somatic cells. The chromosomes in the somatic cells of humans are determined by the fertilization event when a sperm and an egg combine, making a zygote. The zygote has 46 chromosomes in it, 23 from the father and 23 from the mother. Fertilization brings the chromosome number back to what it should be for somatic cells.

The material in this lesson is based on this Meiosis chart. The information was simplified, disassembled, and explained step by step for your student in the lesson. This chart is more detailed and is provided for your information.

Meiosis

Meiosis I	Meiosis II
Paired chromosomes separate	Chromatids separate

Prophase I

1. Chromosomes condense.
2. The nuclear membrane breaks down.

Prophase II

1. Chromosomes DO NOT replicate.
2. Chromosomes begin moving to the middle of the cell.

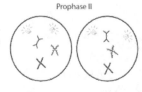

Metaphase I

1. Chromosome pairs line up in the middle of the cell.

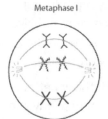

Metaphase II

1. Chromosomes line up in the middle of the cell.

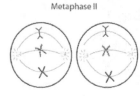

Anaphase I

1. Chromosome pairs separate.

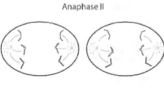

Anaphase II

1. Chromatids separate.
2. the chromatids move toward opposite sides of the cell.

Telophase I

1. Each side of the cell now has a haploid number of chromosomes.
2. The cell begins to divide.

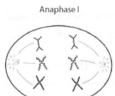

Telophase II

1. Nuclei begin to form around each unpaired set of chromosomes.
2. The cells begin to divide.

Cytokinesis I

1. Two daughter cells are produced.
2. Each have 1/2 the number of chrommomsomes as the parent cell.

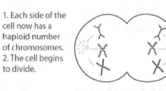

Cytokinesis II

1. There are now four daughter cells.
2. Each daughter cell has a haploid number of chromosomes.

Lab

Meiosis Flip Book

In this lab activity, students create a flip book by drawing the chromosomes into the nuclei for the sequential stages of meiosis. This exercise will be particularly helpful for students who are visual and tactile learners. The somewhat abstract concept of meiosis is reinforced and made more concrete through the drawing and coloring of each picture, and as they riffle through the pages to observe the progression of meiosis from one stage to the next.

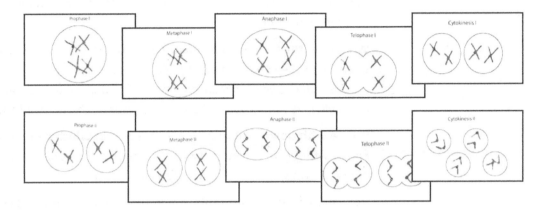

Microscope Lab

Meiosis and the Microscope

The slide for this week is a prepared slide showing cells in different stages of meiosis. This lab provides another opportunity to reinforce the concept of cell division through a visual model. It can be fascinating for students to discover that actual cell division can be seen with the help of a microscope. The cells are fairly easy to see on the prepared slide; deciphering the stages will take some patience.

Your prepared slide will be a cross-section of the anther of a lily flower. The anther houses the flower's pollen sacs and is where pollen is created through cell division, meiosis. The anther cross-section has a butterfly wing shape. Your specimen will have four pollen sacs, one on each "wing."

Focus your lens inside a pollen sac to find the cells undergoing meiosis. Hopefully, your specimen will have all four stages.

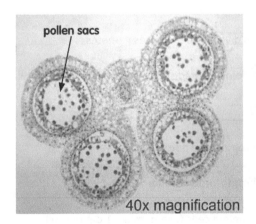

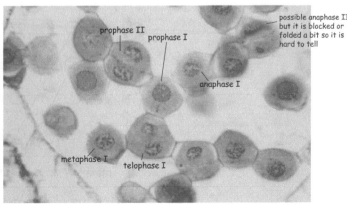

Famous Science Series

Down Syndrome

A trisomy occurs when a diploid organism gets three copies of a chromosome, instead of the two copies it was supposed to get. The symptoms associated with Down syndrome (also known as Down's syndrome) range from mild to severe. This depends on how much of the third chromosome 21 an individual receives. Sometimes it is just a piece of a third chromosome 21, and sometimes it is an entire extra chromosome 21. Down syndrome is one of the few trisomies that result in a viable fetus. Trisomies at the sex chromosomes are also viable. XXX, XXY, and XYY all result in viable fetuses.

What does the term *trisomy* mean? *Sometimes, the chromosome pairs do not separate as they are supposed to. If a pair does not separate, the gamete will have two instead of one of that chromosome. If that gamete goes on to make a person they will have three copies of that chromosome, this is called a trisomy.*

People with Down syndrome have a trisomy at which chromosome number? *People with Down syndrome have three copies of chromosome number 21, instead of two. Down syndrome is also called trisomy 21.*

How many chromosomes total are in the somatic cells of a person with Down syndrome? *46 (23 pairs) + 1 (an extra chromosome number 21) = 47*

What are possible health or developmental impacts for people with Down syndrome? *The range of effects felt by people with Down syndrome is very broad. Some people with Down syndrome have very few health effects and some people have many. Leukemia and other cancers are more common in people with Down syndrome. People with Down syndrome have an increased risk of heart disease. Down syndrome can cause developmental delays as well.*

Show What You Know

Meiosis Divides by Two and Makes You

Match the word with the best definition.

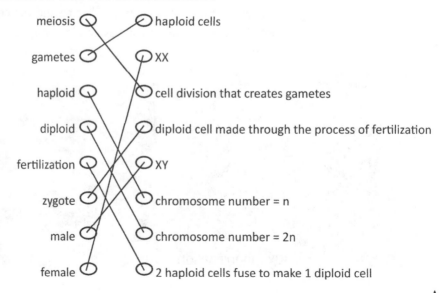

Pandia PRESS

Fill in the Blanks

Mitosis results in *genetically* identical cells. The exact same *genes* and *number* of chromosomes are in the daughter cells as was in the parent cell.

Meiosis results in genetic *variability*. The daughter cells are not genetically *identical* to the parent cell. The chromosome number has been reduced by *half*, which means only *half* the parent's chromosomes are in the daughter cells.

At the start of meiosis I, the parent cell has *2n* chromosomes. During meiosis, the chromosome number is reduced to *n*. The cells that are created during meiosis and cytokinesis are called *gametes*.

During *fertilization*, the two gamete cells (one from a male and one from a female) fuse, making *one* cell, called a *zygote*. In multicellular organisms, the unicellular zygote divides using the process of *mitosis* and *cytokinesis*.

The *23rd* chromosome pair in humans are called the sex chromosomes. In females, this pair is a *homologous* pair. Females have two *X* chromosomes. In males, this pair is not a *homologous* pair of chromosomes. Males have one *X* and one *Y* chromosome. The gender of a person is determined by their *father*. (This always makes me think of Henry VIII. He blamed his wives for his lack of sons.)

Lesson Review

Meiosis Divides by Two and Makes You

All cells come from other cells. For organisms that reproduce sexually, half their DNA comes from their mother and half their DNA comes from their father.

Ploidy = number of sets of chromosomes. n = number of different chromosomes in 1 set. Humans are diploid = 2 sets. In humans, n = 23.

Meiosis and mitosis are both nuclear division. BUT: At the end of meiosis, there are 4 cells with n chromosomes; each cell is a gamete. At the end of mitosis, there are 2 cells with 2n chromosomes; each cell is a somatic cell.

Mitosis	Meiosis
in somatic cells	in gametes
One cell division, resulting in 2 daughter cells.	Two cell divisions, resulting in 4 daughter cells.
The chromosome number stays the same.	The chromosome number is halved going from the parent cell to the 4 daughter cells.
The parent cell and the daughter cells are all genetically identical to each other.	The daughter cells are not genetically identical to each other.
Centromeres divide at anaphase. When the centromeres divide, the sister chromatids separate.	Centromeres do not divide at anaphase I. They divide at anaphase II. When the centromeres divide, the sister chromatids separate.
The daughter cells are diploid. The ploidy of the daughter cells is 2n.	The daughter cells are haploid. The ploidy of the daughter cells is n.

Meiosis

- Meiosis has 2 cycles: meiosis I and meiosis II (PMAT I and PMAT II)
- From prophase I to cytokinesis I it is similar to mitosis, EXCEPT the chromosome pairs separate, not the chromatids.
- At prophase II, the chromosomes in the two cells do NOT duplicate; the rest of the process is similar to mitosis.

Sexual Reproduction

2 ways to say the same thing:

1 male gamete + 1 female gamete + fertilization

= 1st somatic cell (this cell makes an organism with mitosis)

1 sperm + 1 egg + fertilization

= 1 zygote (the zygote makes an organism with mitosis)

crossing over + meiosis + fertilization

= GENETIC VARIABILITY (why you do NOT look exactly like your sister)

The father determines gender. Mothers give an X chromosome to ALL their offspring. Fathers give an X or a Y.

XX = female   XY = male

Teacher Guide

Unit III: Genetics
Chapter 10: Inheritance

Introduction

Chapter 10 is about inherited traits. This subject is a favorite for many students. The concepts of inheritance are fairly straightforward and provide an opportunity for students to apply what they have learned about genetics thus far. But this last chapter on genetics is filled with strange new terminology that might make this seem more confusing than it actually is. Don't shy away from the vocabulary. With continued use, students will pick it up fast.

This is the last chapter in Unit III. There is a Unit III exam that covers the material found in Chapters 7 through 10, in the appendix of the student Workbook. The answer key is found at the end of this chapter.

Preparation for microscope lab: Collect human hair strands in as many different colors as you can. Make sure all but one is untreated. Also, try to get hair samples from someone who is going gray; collect one gray hair and one non-gray hair.

Learning Goals

- Investigate how organisms of the same species come to have unique traits.

- Understand how genotype, phenotype, and traits relate.

- Understand the new terminology and concepts relating to alleles.

- Understand how to use a Punnett square and a probability table.

- Introduce the concept that genes, environment, and choices make you who you are.

- Learn the Law of Independent Assortment of Alleles.

- Learn the Law of Segregation.

- Learn about your family traits.

Extracurricular Resources

Books

Gregor Mendel: The Friar Who Grew Peas, Bardoe, Cheryl

Gregor Mendel: Genetics Pioneer: Life Science, Van Grop, Lynn

Gregor Mendel: And the Roots of Genetics, Edelson, Edward

Online

Visit Pandia Weblinks for videos and websites recommended for this chapter:

www.pandiapress.com/weblinks-biology2

Pandia PRESS

Lesson

What Makes You You?

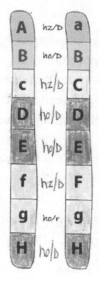

Students are introduced to the new terms *genotype* and *phenotype* in this lesson. Your genes make up your genotype. Phenotype is the expression of your genes, your appearance. Genotype is the main factor affecting phenotype. In other words, the genes an organism has are the main factor affecting their appearance. There are different forms of each gene. Different forms of a gene code for making the same protein; they just make different versions of it. Blond hair versus brown hair, for example. You still have hair; it is just a different color. In this lesson students color alleles as they read.

To help demonstrate how different alleles affect phenotype, I came up with an imaginary creature, qwitekutesnutes [*quite-cute-snoots*]. I used an imaginary creature instead of a real-life organism for several reasons. The genetics I have presented so far in Unit III are straightforward, basic, and simplified. In reality, genetics is often much more complicated. Traits can be controlled by more than two alleles, there can be incomplete dominance, and genes controlling different traits can be very close to each other on a chromosome so that the assortment of the different genes is not entirely independent. Therefore, it would have been close to impossible to be absolutely certain I was getting the genetics completely correct for any living organism. With qwitekutesnutes I had total control of the genetics. I was able to keep it simple, straightforward, and at grade level, while still having an organism to use as an example. Plus qwitekutesnutes are, well . . . quite cute!

Lab

Family Traits

This lab is so much fun. Interview as many close relatives as possible. The answers for some of the traits, such as eye color, are subjective. Let the interviewer decide. If your student is adopted, I recommend you adapt this lab rather than skip it. An adopted child could choose a subject: someone who has access to and knowledge of many blood relatives. The student could gather data and complete the lab based on the subject. Actually, scientists are rarely the subject of their own experiments, so using a third party is a completely legitimate way to conduct this lab.

You may need to help students when they fill in the Family Traits Questionnaire. Answers will vary according to the subjects used. A few notes:

• If everyone interviewed has only the dominant trait, this does NOT mean they are all homozygous for the dominant allele.

• When they occur, students can work back from a homozygous recessive phenotype to determine a heterozygous genotype.

• Blood type is a genetic trait that is not controlled by one dominant or recessive allele. Both alleles are expressed if two different alleles are present. The alleles are co-dominant.

Microscope Lab

Phenotype Under the Scope

This experiment surprised even me with its fun factor. You can see the color molecules along the hair shaft and the parts of the shaft that are unpigmented. Dyed hair looks different from undyed hair. Gray hair has color molecules but they no longer have color in them. Get as many different samples of naturally colored hair as you can. (I asked for samples at a dinner party.) You only need one sample of dyed hair, because they all look the same except coated with a different color.

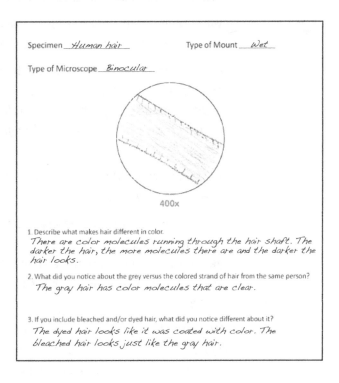

Specimen _Human hair_   Type of Mount _Wet_

Type of Microscope _Binocular_

400x

1. Describe what makes hair different in color.
There are color molecules running through the hair shaft. The darker the hair, the more molecules there are and the darker the hair looks.

2. What did you notice about the grey versus the colored strand of hair from the same person?
The gray hair has color molecules that are clear.

3. If you include bleached and/or dyed hair, what did you notice different about it?
The dyed hair looks like it was coated with color. The bleached hair looks just like the gray hair.

Famous Science Series

Gregor Mendel

There was a lot going on in the field of genetics in the mid to late 1800s. In 1859, Charles Darwin published his seminal work on evolution and natural selection, *The Origin of Species*. In 1866, Gregor Mendel published the results of seven years of research that examined how traits were passed from parents to their offspring. Unfortunately, Mendel published in an obscure Austrian periodical and Darwin never learned of Mendel's research. It was unfortunate because although Darwin was able to explain the "what" of evolution, he did not know the mechanism that explained the "how" of evolution. It was, at that time, a serious flaw to his theory. Evolution is covered in more detail in Unit 5.

If possible, do the Famous Science Series for chapter 10 and the "Make Your Own Qwitekutesnute" Activity together.

When and where was Gregor Mendel born? *Heinzendorf, Austria, on June 22, 1822. Heinzendorf, Austria, is now Hyncice, Czechoslovakia.*

Pandia PRESS

What did Mendel do so that he could continue his education? *He joined the Augustinian Abbey of St. Thomas in what is now Brno, Czechoslovakia, in 1843.*

What is the blending theory of inheritance? *The blending theory states that the traits of an individual result from a blending of the traits of the parents. A tall parent and a short parent would have children of medium height.*

From 1856 to 1863, Mendel conducted an experiment with over 28,000 plants. What type of plant did he use? *Pea plants, common garden pea*

What seven traits did he study in these plants? *flower color, flower position, stem length, seed shape, seed color, pod shape, and pod color*

Did Mendel prove or disprove the blending theory of inheritance? *When Mendel crossed plants with specific traits, he got one of the two traits, not something in the middle. This disproved the blending theory of inheritance. From this he concluded that traits are passed on unchanged from parents to offspring by "units," now called genes.*

What two laws did Mendel discover? *Law of Independent Assortment of Alleles and the Law of Segregation*

When Mendel crossed true-breeding green peas and white peas, he got all green peas in the F1 generation. When he crossed two of the green peas from the F1 generation he got ¾ green peas and ¼ yellow peas in the F2 generation. Which is the dominant trait and which is the recessive trait? *Green is dominant, yellow is recessive.*

Was Mendel famous in his lifetime? *No, the scientific community took little notice of his work. It was not until 1900, sixteen years after his death, that three different scientists working to explain the laws of inheritance rediscovered Mendel's work.*

Activity

Make Your Own Qwitekutesnute

There is no answer key for this activity. Drawings will vary depending on the traits of the qwitekutesnute.

This activity is designed to be done directly following Famous Science Series. You might need to remind your student that the assortment of alleles is random. You do not always get what you want. This might seems like silly advice, but in my experience, students often want certain traits for their qwitekutesnute, which can affect the randomness of the traits the qwitekutesnute gets.

Show What You Know

Inheritance

1. Match the word with the best definition.

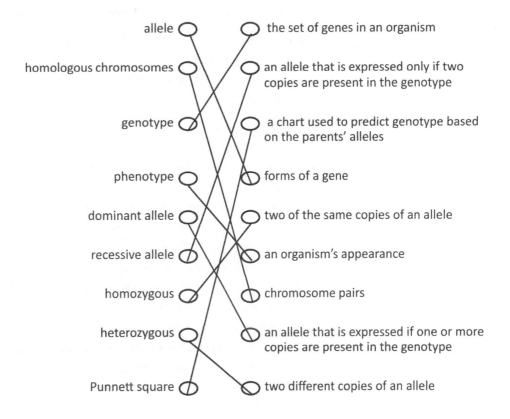

2.

	H	h
h	Hh	hh
h	Hh	hh

H = spiked hair   h = ear tufts

Genotype	Genotype Probability	Genotype Fraction	Genotype Percentage	Phenotype	Phenotype Probability	Phenotype Fraction	Phenotype Percentage
Hh	2:4	2/4	50%	Spiked hair	2:4	2/4	50%
hh	2:4	2/4	50%	Ear Tufts	2:4	2/4	50%

What is the probability of a qwitekutesnute baby from this pair having ear tufts?
2:4 = 50%

If the qwitekutesnute parents have 12 babies, how many should have ear tufts?
Will that many definitely have ear tufts? *Six. No, this is just the likelihood, the probability.*

If qwitekutesnute parents both have gray eyes (a dominant trait among qwitekutesnutes), could they have green-eyed offspring (a recessive trait)? Explain your answer. *Yes, if both parents are heterozygous, Ee, for eye color, they could have a baby that was homozygous recessive for eye color.*

If qwitekutesnute parents both have 4 whiskers, a recessive trait, could they have offspring with 7 whiskers? Explain your answer. *No, both the parents are homozygous recessive for the trait so neither parent has the dominant allele to pass on to their offspring.*

Multiple Choice

1. Law of Segregation states *allele pairs separate during meiosis*
2. Law of Independent Assortment states *allele pairs assort independently of one another*
3. The scar on your chin is an example of *phenotype*
4. The allele pair Ww is *heterozygous*
5. The allele pair BB is *homozygous dominant*
6. The allele pair ee is *homozygous recessive*
7. If two parents with brown hair have a baby with blond hair, the allele for blond hair must be *recessive*
8. Traits are *inherited and acquired*
9. Your genotype is *the set of genes in the somatic cells in your body*
10. Your traits are your *phenotype*

11. Extra Practice

HH x HH phenotype: 100% spiked hair

	H	H
H	HH	HH
H	HH	HH

HH x Hh phenotype: 100% spiked hair

	H	H
H	HH	HH
h	Hh	Hh

HH x hh phenotype: 100% spiked hair

	H	H
h	Hh	Hh
h	Hh	Hh

Hh x Hh phenotype: 75% spiked hair, 25% ear tufts

	H	h
H	HH	Hh
h	Hh	hh

Hh x hh phenotype: 50% spiked hair, 50% ear tufts

	H	h
h	Hh	hh
h	Hh	hh

hh x hh phenotype: 100% ear tufts

	h	h
h	hh	hh
h	hh	hh

Lesson Review

Inheritance

Traits = inherited and acquired characteristics
Inherited characteristics = traits determined by genes, e.g. eye color, heart murmur
Acquired characteristics = traits from life experiences, ex. scars

Meiosis separates alleles so there is one copy of an allele in a gamete.
Fertilization = 1 haploid set of mother's chromosomes + 1 haploid set of father's

Mitosis – This cell replicates to make more cells that are all genetically identical to the first cell, the zygote.

The chromosomes of diploid organisms come in pairs of homologous chromosomes.
One diploid set of offspring's chromosomes = offspring's genotype

Genotype determines phenotype.
Genotype = a set of genes in an organism
Phenotype = the appearance of the organism

Alleles are the forms a gene comes in.
Alleles can be dominant or recessive.
Allele vocabulary and rules:
- An uppercase letter = a dominant allele
- A lowercase letter = a recessive allele
- WW = homozygous dominant
- Ww = heterozygous
- ww = homozygous recessive

Your inherited traits are determined by the alleles you inherit.

Law of Segregation = allele pairs separate during meiosis
Law of Independent Assortment = allele pairs separate independently of each other

Teacher Guide

Unit III: Genetics
Answer Key Unit Exam Chapters 7–10

The exam for Unit III is found in the appendix of the student Workbook.

1. Vocabulary (2 points each, 18 points total)

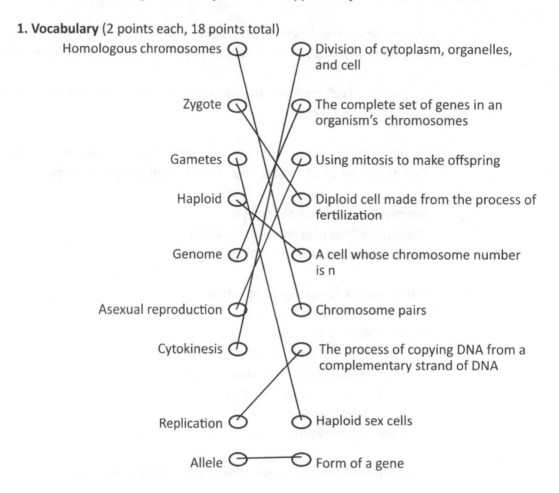

Homologous chromosomes — Division of cytoplasm, organelles, and cell

Zygote — The complete set of genes in an organism's chromosomes

Gametes — Using mitosis to make offspring

Haploid — Diploid cell made from the process of fertilization

Genome — A cell whose chromosome number is n

Asexual reproduction — Chromosome pairs

Cytokinesis — The process of copying DNA from a complementary strand of DNA

Replication — Haploid sex cells

Allele — Form of a gene

2. True or False (1.5 points each, 15 points total. Plus 1 extra credit each for correcting the false statements.)

__F__ ~~Cytokinesis~~ is the part of the cell cycle when proteins are made. **Interphase**

__T__ In humans, it is the gamete from the father that determines gender.

__F__ During translation, RNA gives the information for building a protein to ~~DNA~~. **ribosomes**

__F__ ~~Meiosis~~ results in genetically identical cells. **Mitosis**

__T__ DNA contains all the information needed to run your body.

__T__ If an organism has 10 chromosomes in their somatic cells, their gametes will have 5 chromosomes in them.

__F__ The genes you inherit are responsible for ~~all your traits~~. **some of your traits. Other traits, such as scars, are acquired.**

__T__ The specific bonding of base pairs makes the replication of DNA strands possible.

__F__ Meiosis starts with 1 cell and ends with ~~2~~ cells. **4**

__T__ A diploid cell has 2n chromosomes in it.

3. Multiple Choice (2 points each, 20 points total)

The acronym for remembering the steps mitosis and meiosis follow is <u>PMAT</u>.

The gender of humans is determined by the 23rd chromosome in the following pattern: <u>male XY, female XX</u>.

During prophase, the _____ become visible. <u>homologous chromosomes</u>

At the beginning of meiosis there is/are _____ chromosome(s). At the end of meiosis there are _____ chromosomes. <u>2n, n</u>

The three parts of the somatic cell cycle are <u>interphase, mitosis, cytokinesis.</u>

DNA is in the shape of a <u>double helix</u>.

During fertilization, two gametes fuse to make a <u>zygote</u>.

How many chromosomes are in a human skin cell? <u>46</u>

The correct order going from the smallest building block to the largest is <u>nucleotide base, codon, gene, chromosome, homologous chromosome, genome</u>

When bacteria reproduce using asexual reproduction, it is called <u>binary fission.</u>

4. Written Answers.

Describe how a protein is made. Start your description of the events inside the nucleus. Make sure to use the correct vocabulary terms when describing the process of protein synthesis. (10 points total, 2 for each of the four steps and 1 point for the correct use of the terms *translation* and *transcription*.)

1. The gene that codes for the making of that protein unravels.

2. An RNA molecule is built along a gene sequence of DNA. This is called transcription.

3. The RNA molecule (the RNA template) leaves the nucleus.

4. The RNA molecule attaches to a ribosome. The transfer of information from an RNA molecule to a ribosome is called translation. Each codon on the RNA molecule codes for the synthesis of one amino acid. The entire RNA molecule codes for the synthesis of a protein.

Draw a complementary strand of DNA above this sequence, and draw a complementary strand of RNA below this sequence. (2 points each, 4 points total)

T A G C C A A T C G A T C G G

A T C G G T T A G C T A G C C

U A G C C A A U C G A U C G G

What is the name of the process where DNA makes a copy along a strand? (1 point) <u>Replication</u>

List two things that are different between mitosis and meiosis. (1.5 points each, 3 points total. There are more than two things; you can give extra credit for listing more.)

1. At the end of mitosis there are 2 cells that are genetically identical to each other and to the parent cell.
2. At the end of meiosis there are 4 cells that are not genetically identical to each other or to the parent cell.
3. Mitosis goes through 1 divisionary cell cycle from start to finish, and meiosis goes through 2.
4. Meiosis: starts 2n (diploid) → ends n (haploid). Mitosis: starts as 2n and ends as 2n.
5. Mitosis occurs in somatic cells and meiosis occurs in sex cells.

List two things that are the same for mitosis and meiosis. (1.5 points each, 3 points total)
1. Both are used by the cells of sexually reproducing organisms to divide nuclei.
2. Both involve the division of chromosomes.

Put these in correct order using numbers from 1 to 4, 1 being the first stage of mitosis and 4 being the last stage of mitosis. (1 point each, 4 points total)

__4__   __2__   __1__   __3__

5. Punnett Square (5 points each table, 10 points total)

You are so lucky. Your parents bought you two qwitekutesnutes for your birthday! The pet shop owner sold your mom two females. At least that's what he told her. Except that now it looks like one of them is going to have babies. You are really excited, and your mom is not. I wonder why? One of the qwitekutesnutes has 3 toes (tt) and one has 5 toes (Tt). If your qwitekutesnutes have 4 babies, how many should have 3 toes? How many should have 5 toes? To answer these questions, fill in the Punnett square. The allele for 3 toes is t. The allele for 5 toes is T.

	t	t
T	Tt	Tt
t	tt	tt

Fill in the probability table using the data from the Punnett square.

Genotype	Probability	Fraction	Percentage
Tt	2 in 4	2/4 or 1/2	50%
tt	2 in 4	2/4 or 1/2	50%

Phenotype	Probability	Fraction	Percentage
5 toes	2 in 4	2/4 or 1/2	50%
3 toes	2 in 4	2/4 or 1/2	50%

How many babies should have 5 toes? (2 points) <u>They have 4 babies, therefore 2 are predicted to have 5 toes.</u>

How many babies should have 3 toes? (2 points) <u>They have 4 babies, therefore 2 are predicted to have 3 toes.</u>

When they are born, three babies have 3 toes and one has 5 toes. How do you explain this? (4 points)
<u>The probability just gives possible predictions. It does not tell you what will definitely happen.</u>

Qwitekutesnutes have 3 toes if their genotype is tt. What is this type of allele, t, called? (2 points) <u>recessive</u>

Qwitekutesnutes have 5 toes if their genotype is TT or Tt. What is this type of allele, T, called? (2 points) <u>dominant</u>

(____ /100) x 100 = _____ + _____ extra credit points = _____

Unit IV: Anatomy and Physiology
Chapter 11: Multicellular Organisms

WEEKLY SCHEDULE

Two Days

Day 1
- ❏ Lesson
- ❏ Dissection Lab

Day 2
- ❏ FSS
- ❏ Lesson Review
- ❏ SWYK

Three Days

Day 1
- ❏ Lesson

Day 2
- ❏ Dissection Lab

Day 3
- ❏ FSS
- ❏ Lesson Review
- ❏ SWYK

Five Days

Day 1
- ❏ Lesson

Day 2
- ❏ Dissection Lab

Day 3
- ❏ FSS

Day 4
- ❏ Lesson Review

Day 5
- ❏ SWYK

FSS: Famous Science Series
MSLab: Microscope Lab
SWYK: Show What You Know

Introduction

Units II and III focused on the cell. While the cell is the basic structural and functional unit for multicellular organisms, it isn't the sum total. Unit IV is the anatomy and physiology unit. This is where we look at how all those specialized cells come together to make a functional multicellular organism. This unit covers the major organs and organ systems in plants and humans. This is the longest unit in the book. Chapter 11 is an introductory chapter. It examines how one cell divides and specializes into different tissues and organs to become an organism. Chapter 12 covers the major organs and organ systems of plants. Chapter 13 explains plant reproduction in angiosperms, the most common type of plant on Earth today. Chapters 14 through 19 discuss two to three major organ systems or related anatomy each week:

Chapter 14: The nervous system and the senses.

Chapter 15: The integumentary system and the digestive and urinary systems.

Chapter 16: The endocrine system and the reproductive system.

Chapter 17: The circulatory system and the respiratory system.

Chapter 18: The skeletal system and the muscular system.

Chapter 19: The immune and lymphatic systems as a pair. This chapter ends with a section called Putting It All Together. Students investigate how all these organ systems must work together to make a functioning human.

Throughout Unit IV, students label anatomy and physiology systems by coloring directly in their Workbooks. When students create their own reference material, they are gaining "ownership" over their material. Do not look at these as just coloring pages. When students interact with the material through labs, in this case coloring, they are creating a different dynamic and relationship to the material. Many educators feel that creating ownership over academic material is a powerful method and tool to use when teaching. I use it primarily when teaching things students cannot see, such as internal organs, DNA, and the parts of an atom (from RSO Chemistry 1).

Learning Goals

- Recognize that in multicellular organisms, cells make tissue, tissues makes organs, organs make organ systems, and organ systems make organisms.

- Learn about the four main types of tissues.

- Understand that the failure of one organ can result in the death of an organism.

- Learn how to perform a dissection.

- Learn how to make microscope slides as a part of a dissection.

- Investigate the organ systems common to animals by dissecting a frog.

Extracurricular Resources

Books

Organ Donation Risks, Rewards, and Research, Brezina, Corona

Super-Flea and Other Animal Champions: Cells, Tissues, and Organs, Spilsbury, Richard

Human Organs, Lew, Kristi

Organ Transplants, Campbell, Andrew

Alexander Fleming and the Story of Penicillin, Bankston, John

Discovery of Penicillin, de la Bedoyere, Guy

Alexander Fleming: The Man Who Discovered Penicillin, Tocci, Salvatore

Online

Visit Pandia Weblinks for videos and websites recommended for this chapter:
www.pandiapress.com/weblinks-biology2

Lesson

The House that Cells Built

After the chapters in Unit 3, the text for this chapter is going to seem easy. Why not couple it with the old black-and-white video of Frankenstein mentioned above, or choose another version.

Dissection Lab

Inside View of a Frog

Warnings about this lab: When ordering a frog, call and talk to someone. Make sure they will take the frog back if the frog is not of good quality.

This lab requires teacher involvement ensuring students do not cut or puncture themselves.

The general lab and microscope lab are combined for this chapter. Have students answer the questions on each lab sheet BEFORE going on to the next part.

This is the best lab for looking at how organs and organs systems fit in an organism. If you and your student are wavering about performing this lab, remember, there is no virtual substitute for the real thing. However, a good *introduction* for this lab is a virtual frog dissection. A virtual dissection shows proper cutting technique. It also shows the position of some, but not all, of the organs students will look for during the real dissection. But not every student is comfortable with something having been killed so they can study how organs fit together inside of an organism. If that is the case, you can try using one of the virtual sites in substitution of the frog specimen.

Possible Answers

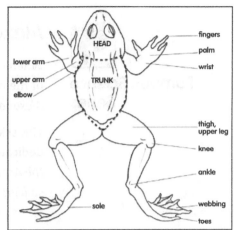

1. How many fingers does the frog have? *4 fingers*

2. One method for determining gender in a frog is to examine the fingers on its hands. Male frogs have thick pads on their "thumbs." Female frogs do not. Does your frog have these pads? Do you think your frog is a male or a female? *Answers will vary depending on the sex of the frog.*

3. How many toes does the frog have? *5 toes*

4. What is the length of the frog? *Answers will vary*

5. Describe the color difference between the top of the frog and the belly of the frog. The color difference protects a frog from predators when it is in the water. How do you think this works? *The frog is dark on top and light on its belly. When predators look down from above, the dark color makes the frog blend in with the water, which also appears dark from above. When predators look up from below, the light color makes the frog blend in with the sunlight shining down on the water, which appears light from below.*

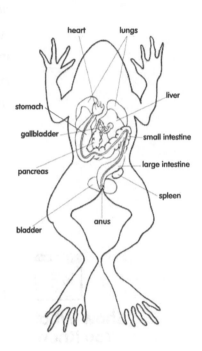

6. A frog's nares are on the top of its nose. How would this help the frog when swimming? *The position of the nares allows the frog to be underwater except for its nares. This is better for swimming because most of the frog's body can be submerged.*

7. How do you think a frog's webbed feet help it in the water? *The webbing between the toes makes frogs more "aqua"dynamic through the water, allowing for more push each time the frog moves its hands and feet while swimming.*

8. The outer covering of the frog, its skin, is its largest organ. What do you think your largest organ is? *Skin*

9. What happened when you put a probe through the frog's nostril? *The probe comes out through the internal nostril.*

10. Live frogs have sticky tongues. How do you think this helps them catch their dinner? *A frog catches a bug on its tongue so the bug cannot escape.*

11. Eustachian tubes equalize pressure in the ear. What external organ do you think they work with? *Tympanic membrane*

12. Think about experiences you have had with drainage between your mouth and nose. Do you think that your nasal passage goes directly into your mouth too? *Answers might vary, but yes*

13. Look at the blood vessels under the skin. What is their purpose? *To circulate blood, which carries oxygen, water, and food throughout the frog's body and to carry waste out of it*

14. The organs in the digestive system are the mouth, tongue, teeth, esophagus, stomach, small intestine, large intestine, and anus. If this frog ate a bug, write the path it would take. *Into mouth → through teeth → then tongue → esophagus → stomach → small intestine → large intestine → anus*

Famous Science Series

Alexander Fleming

When were antibiotics discovered? Who discovered them? What was the name of the first antibiotic? *The first antibiotic to be discovered was penicillin in 1928, by Alexander Fleming.*

This scientist fought in what war? What did he see in that war that caused him to dedicate his life to finding an antibacterial agent? *Fleming served as a captain in the Army Medical Corps throughout World War I. During the war, he witnessed many soldiers die from infected wounds. When the war was over, he dedicated himself to finding an antibacterial agent.*

What percentage of Union soldiers who died, died from infectious disease? *70 percent*

What is the estimated percentage of Confederate soldiers who died from infectious disease? *75 percent*

Of the 620,000 soldiers who died in the Civil War, approximately how many died from infectious disease? *414,000*

Show What You Know

Multicellular Organisms

Multiple Choice

1. Your blood connects your entire body. Even though it is a liquid it is considered a type of this tissue: *Connective tissue*
2. Your brain has sensors going back and forth to it as it constantly monitors your environment. It relies on this type of tissue to transmit information: *Nerve tissue*
3. Your body's largest organ is your skin. Its primary job is to protect the inside from the outside and to hold you together. Your skin is lined with this type of tissue: *Epithelial tissue*
4. Your heart beats your entire life. It never stops working. It relies on this tissue to keep it beating: *Muscle tissue*

Matching

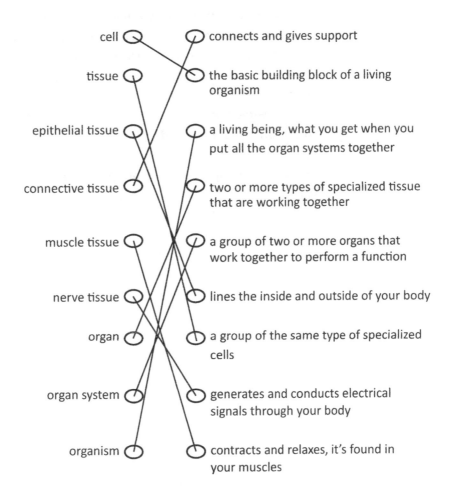

cell	connects and gives support
tissue	the basic building block of a living organism
epithelial tissue	a living being, what you get when you put all the organ systems together
connective tissue	two or more types of specialized tissue that are working together
muscle tissue	a group of two or more organs that work together to perform a function
nerve tissue	lines the inside and outside of your body
organ	a group of the same type of specialized cells
organ system	generates and conducts electrical signals through your body
organism	contracts and relaxes, it's found in your muscles

Lesson Review

Multicellular Organisms

For multicellular organisms:

After fertilization there is one cell = zygote.

That cell divides to make more cells that are genetically identical to it.

Those cells begin to specialize = some group together to make a heart, some group together to make a brain.

The brain cells are genetically identical to the heart cells.

The brain cells and the heart cells have different parts of their genes turned on.

The cells make tissues:

2 or more types of tissue make an organ.

2 or more types of organ make an organ system.

Organ systems group together to make an organism.

There are four main types of tissue:

1. **Epithelial** = lines inside and outside of your body – the cheek cells you stained and looked at with your microscope in chapter 3 were from epithelial tissue.

2. **Connective** = connects and gives support – blood is connective tissue, so is the cartilage you see when you separate a chicken leg from the thigh.

3. **Muscular** = contracts and relaxes – tighten the muscles in your arms or legs and feel the muscle tissue contract and relax.

4. **Nerve tissue** = generates and conducts electrical signals to send messages to and from your brain and the rest of your body. For example, when you bite your lip, signals are sent via your nerve tissue to your brain saying that pressure is being applied to your lip. Your brain interprets the signals and you "feel" the pressure on your lip.

Unit IV: Anatomy and Physiology
Chapter 12: Plant Anatomy

FSS: Famous Science Series
MSLab: Microscope Lab
SWYK: Show What You Know

Introduction

The information in Chapter 12 refers to angiosperms, the most common type of plant. Students dissect a plant in this chapter. They make slides while they are dissecting, improving their dissecting technique and their slide-making technique.

Learning Goals

- Learn the names and functions of the three main tissue types in plants.

- Learn about the two primary organ systems of plants: the root and shoot systems.

- Learn the names and functions of the main organs in plants.

- Learn the concept of homeostasis.

- Learn about translocation and transpiration in plants.

Extracurricular Resources

Books

The Flowering Plant Division, Stefoff, Rebecca

Plant Parts (the Life of plants), Spilsbury, Louise, Spilsbury, Richard

Online

Visit Pandia Weblinks for videos and websites recommended for this chapter:

www.pandiapress.com/weblinks-biology2

Lesson

Plants, Roots, and Shoots

The number of organs and organ systems in plants are fewer than in people. It is the reason I put the chapters on plant anatomy and physiology before humans. That way, students can get a feel for how the organs and the organ systems, even though separate, must all work together to have a healthy functioning organism.

As students will learn in Chapter 30, not all plants have vascular tissue. If you would like to further investigate and look at vascular tissue in plants, the following experiment is a good one for that: Split the stem of a white or light-colored carnation up toward the flower 15 to 20 cm. Put one side of the split stem in a glass of clear tap water. Put the other side of the stem in a glass of water with a liberal amount of blue food color in it. Make sure both sides of the stem stays immersed in water until the blue color reaches the flower. When this happens, take the carnation out of water and cut the stem above the water line; split it further up if you want. Students will be able to see the straw-like tubes of the xylem and phloem.

Homeostasis is an important concept. I introduce it here in a situation that is easy to understand. The topic of homeostasis is discussed several more times, as it relates to humans, throughout the rest of Unit IV. Go over the illustration showing transpiration and make sure students understand how it works.

Dissection and Microscope Lab

Plants: The Inside Story

The dissection lab and microscope lab are combined in this chapter. This lab requires teacher involvement. Dissections are a multistep process, where students are asked to look for little parts and details. It might be necessary to help with some of the cutting and slicing. The primary root in particular is hard to cut.

Be especially careful when pulling the plant from the soil. If you are careful, you will be able to see a root cap and root hairs with the microscope. Make sure as much dirt as possible is removed from the roots; again be very gentle. Use water and then let the plant dry. You might choose to clean the plant first thing in the morning, lay it on a towel to dry, and then do the dissection when the plant is dry. If you do this, check on the plant periodically to make sure it does not start to wilt. Wilting makes it harder to get good slices.

Remember: When making microscope slides, the thinner the slice the better.

Famous Science Series

Isabella Abbott

I chose an organism that wasn't a plant and a botanist who doesn't study plants for this Famous Science Series. Seaweed looks like a plant, and many people call it one, but it does not have a root and shoot system like plants do. Mosses, which are plants, do not have vascular tissue either.

What is Isabella Abbott's given name? What does her given name mean? When and where was she born? *Isabella Abbott's given name was Isabella Kauakea Yau Yung Aiona. It means "white rain of Hana." She was born in Hana, Maui, Hawaii, on June 20, 1919.*

She was the first native Hawaiian woman to receive what college degree? *A Ph.D. in science*

In 1982, Abbott was hired by the University of Hawaii to study ethnobotany. What does the term *ethnobotany* mean? *Ethnobotany is the study of the interaction between plants and humans.*

Abbott was one of the the world's leading experts on what type of organism? Many people think this organism is a plant. It isn't. What type of organism is it? *Seaweed. It is a protist.*

Why isn't it a plant? *(The top three answers are all you need, inclusion of the last two deserve extra credit.)*
Seaweed doesn't have roots. Seaweed uses holdfasts to anchor them.
Seaweed doesn't have stems or leaves like plants.
Seaweed doesn't have vascular tissue.
The cell walls of seaweed are made from a different type of molecule than the cell walls of plants.
Seaweed uses a different type/form of chlorophyll molecule for photosynthesis.

Show What You Know

Plant Anatomy

1. Match the vocabulary word on the left side with the definition on the right side.

dermal tissue	primary root, lateral root, root hairs
vascular tissue	leaves, buds, stems, flowers
ground tissue	covers and protects the outside of plants
root system	tiny cells where water is absorbed from the soil
shoot system	shoots, leaves, and flowers grow from them
primary root	protects the root tip as it grows through the ground
lateral roots	provide support and allows for the flow of water and nutrients
leaves	reproductive organ in plants
root hairs	organ where photosynthesis takes place
root cap	large root with smaller ones growing out of it
buds	transports food, water, and nutrients throughout the plant
stems	small roots growing out of primary root
flower	what most of the plant is made from

2. Label the parts of the plant.

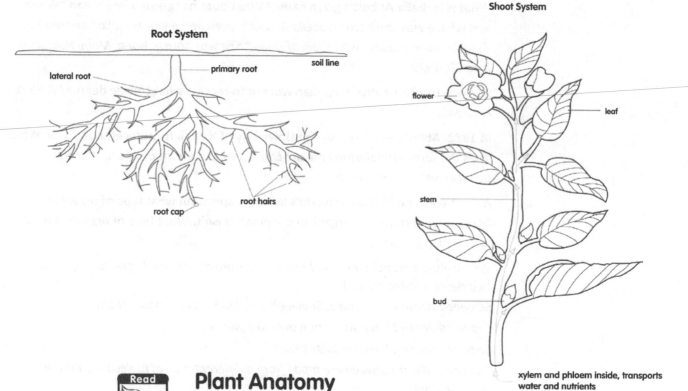

Root System

soil line

lateral root

primary root

root hairs

root cap

Shoot System

flower

leaf

stem

bud

xylem and phloem inside, transports water and nutrients

Read

Lesson Review

Plant Anatomy

Plants have three types of tissue:

1. Dermal tissue–protects and covers the outside of plant, like your skin
2. Vascular tissue = xylem and phloem–transports food, water, and nutrients
3. Ground tissue–makes up most of plant = inside part of roots, leaves, tree trunks, stem (where photosynthesis occurs)

The root system is below the ground. The parts:

1. Primary root–thickest part of root
2. Lateral root–grows from primary root
3. Root cap–protects tip of root as it pushes through the ground
4. Root hairs–one cell thick, water and nutrients are absorbed through these

The shoot system is above the ground. The parts:

1. Leaves–site of photosynthesis
2. Stems–supports the plant, site of translocation
3. Buds–leaves, branches, and flowers grow from buds
4. Flower–reproductive organ of plants

Homeostasis = maintain internal condition within a normal range

Translocation = the movement of food from leaves, where it is made, to other parts in the plant where it is stored or used

Plants use transpiration to maintain water homeostasis:

1. Water evaporates from leaf or is used during photosynthesis → concentration of water in leaf drops
2. The leaf needs water → roots absorb water
3. Water travels up through xylem in the stem to the leaf → the concentration of water in the leaf is now within the normal range, the plant has maintained water homeostasis.

Unit IV: Anatomy and Physiology
Chapter 13: Plant Reproduction

WEEKLY SCHEDULE

Two Days

Day 1
- ❑ Lesson
- ❑ Dissection Lab

Day 2
- ❑ FSS
- ❑ Lesson Review
- ❑ SWYK

Three Days

Day 1
- ❑ Lesson
- ❑ Dissection Lab

Day 2
- ❑ FSS

Day 3
- ❑ Lesson Review
- ❑ SWYK

Five Days

Day 1
- ❑ Lesson

Day 2
- ❑ Dissection Lab

Day 3
- ❑ FSS

Day 4
- ❑ Lesson Review

Day 5
- ❑ SWYK

Introduction

Chapter 13 looks at the parts of the flower. The flower is the reproductive organ of flowering plants called angiosperms. Angiosperms are the most common type of plant on Earth today. The reproductive strategies used by the other three main divisions of plants are discussed in less detail.

Learning Goals

- Learn the names and functions of the parts of a flower.

- Learn more about the process of fertilization.

- Learn the purpose and function of seeds and fruits.

- Learn how a plant goes from a zygote to a germinated plant.

- Learn about the reproductive structure and method of reproduction of gymnosperms.

- Learn about the reproductive structure and method of reproduction of ferns and mosses.

Extracurricular Resources

Book

Flowers (Plant Facts), McEvoy, Paul

Online

Visit Pandia Weblinks for videos and websites recommended for this chapter:

www.pandiapress.com/weblinks-biology2

FSS: Famous Science Series
MSLab: Microscope Lab
SWYK: Show What You Know

Lesson

Making More Plants

The flower is the reproductive organ of angiosperms. For sexually reproducing organisms, fertilization occurs when sperm meets egg and a zygote forms. This is true for plants and for humans. The reproduction of humans is covered in Chapter 16. There are commonalities between the terms and processes for all sexually reproducing multicellular organisms.

This illustration of the flower and the illustration of the cherry tree show slightly different ovaries. Both types are common representations. *Carpel* is the British term for pistil.

Dissection Lab

Flower and Seed: Inside View

The general lab and microscope lab are again combined for this chapter. Students dissect a flower and a seed, taping parts to their lab sheet and labeling. Images of a flower and seed are already labeled on the lab sheet. A gentle touch, patience, and attention to detail are required to locate and dissect each part. Be careful with the ovary and ovules; they are particularly delicate.

There are two main classes of angiosperms: monocotyledons and dicotyledons. These two have different types of leaves and seeds. Monocotyledons have long, parallel veins. Lilies and grasses are monocotyledons. Dicotyledons have complex, net-veined leaves. Maples and most other angiosperms are dicotyledons. A lima bean is the seed of a dicotyledon. For more information on the differences between the two types of angiosperms, refer to Chapter 30.

Famous Science Series

Sunflower

Where did sunflowers originate? *Sunflowers are native to the Americas. Native Americans have used sunflowers for thousands of years.*

Sunflowers have been found at archaeological sites. How old are the remains? *The remains of sunflowers have been found at an archeological site in North America dating from 3000 BCE. There is evidence that Native Americans began cultivating sunflowers by at least 2300 BCE.*

Today, what country is the number one consumer of sunflowers? *Russia*

Sunflowers became important in this country because of two religious holidays. Name the country, the holidays, and why the sunflower is important to this country. *During Lent and Advent, the Russian Orthodox Church did not allow many foods that were rich in oil to be used. When sunflowers came to Russia, they could be used as an oil source during these holy days.*

The rulers of this country sent soldiers into battle with packages of sunflower seeds in what quantity? *The Russian Czars sent a two-pound pack of sunflower seeds with soldiers.*

In 1986, workers at the Chernobyl nuclear power plant caused an explosion that released massive amounts of radioactive material into the surrounding environment. How were sunflowers used to help clean up this problem? *Sunflowers absorb toxic waste from soil and water. Rafts of floating sunflowers were used to extract 95 percent of the radioactive material that was in water.*

Show What You Know

Plant Reproduction

1. Match the vocabulary word with its definition.

pistil—female reproductive structure of an angiosperm

seed—survival capsules, seed coat on outside and food inside

petals—the part of an angiosperm that attracts pollinators

stamen—male reproductive structure of an angiosperm

anther—where pollen is produced for an angiosperm

pollen—male gametes

ovules—female gametes

flower—reproductive organ of angiosperms

angiosperm—flowering plant

gymnosperm—seed-producing plants that do not produce fruit

cones—reproductive organs of gymnosperms

style—part of the pistil that joins the stigma and the ovary

fruit—protects seeds and aids in their dispersal

ovary—where the ovules are stored in an angiosperm

filament—the part of the stamen that supports the anther

spore—used by mosses and ferns in reproduction

stigma—top part of the pistil where pollen is trapped

pollen tube—tube that grows through the stigma to the ovary

germination—when a seed first begins to grow

2. Label the flower.

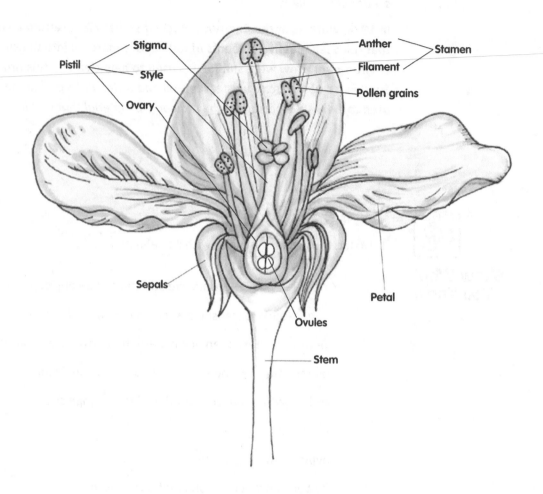

Stigma
Anther
Stamen
Pistil
Filament
Style
Pollen grains
Ovary
Sepals
Petal
Ovules
Stem

Lesson Review

Plant Reproduction

When you are going over the review, use the Socratic method and use this material to ask students questions. For example, when you get to the parts of a flower, you can ask what they are and what they do. Only give answers if students need help.

This is a good time to review meiosis:

Sexually reproducing organisms use the process of meiosis to make gametes.

Gametes have one set of chromosomes in them. They are haploid = n

Male gametes are called sperm in people. In most plants they are called pollen.

Female gametes are called eggs in people, and ovules in most plants.

When the gametes fuse during fertilization, they make a zygote, which has two sets of chromosomes. The zygote is diploid = 2n.

The zygote divides using mitosis. This is how the plant grows. The cells resulting from mitosis are genetically identical.

Angiosperms = flowering plants (even if there is not an obvious flower – such as with grasses)

Parts of a flower:

flower = reproductive organ of angiosperms

pistil = female reproductive organ of angiosperms = stigma, style, ovary

stigma = pollen sticks to stigma, then travels down through the style, creating a pollen tube

style = goes from the stigma to ovary

ovary = where the female gametes (ovules) are and where fertilization occurs in angiosperms

stamen = male reproductive organ of angiosperms

anther = where male gamete (pollen) is produced in angiosperms

filament = the stalk that supports the anther

sepals = protect buds as flower is growing

petals = attract pollinators

seed = where zygote develops into an embryo = a survival capsule that protects and nourishes the embryo

fruits = help with the dispersal of seeds

germinate = when the embryo breaks out of the seed coat and begins to grow

Review the circular process of reproduction using the cherry tree illustration in the student textbook.

Gymnosperms:
- plants that use cones to reproduce (they do NOT make flowers or fruit)
- cones = the reproductive organ of a gymnosperm
- have male cones that make pollen and female cones with female gametes where fertilization takes place
- reproduce sexually
- use wind to disperse pollen from male to female cones

Mosses and ferns are plants that use spores instead of seeds to reproduce.

Unit IV: Anatomy and Physiology
Chapter 14: Nervous and Sensory Systems

WEEKLY SCHEDULE

Two Days

Day 1
- ☐ Lesson 1 & Activity
- ☐ FSS 1
- ☐ Lab 1

Day 2
- ☐ Lesson 2 & Activity
- ☐ Lab 2
- ☐ FSS 2
- ☐ Lesson Review
- ☐ SWYK

Three Days

Day 1
- ☐ Lesson 1 & Activity
- ☐ FSS1
- ☐ Lab 1

Day 2
- ☐ Lesson 2 & Activity
- ☐ Lab 2
- ☐ FSS 2

Day 3
- ☐ Lesson Review
- ☐ SWYK

Five Days

Day 1
- ☐ Lesson 1 & Activity

Day 2
- ☐ FSS 1
- ☐ Lab 1

Day 3
- ☐ Lesson 2 & Activity

Day 4
- ☐ Lab 2
- ☐ FSS 2

Day 5
- ☐ Lesson Review
- ☐ SWYK

Introduction

Chapters 14 through 19 cover human anatomy and physiology. Each chapter teaches at least two organ systems and is split into two sections. Each section has a lab (or labs), a Famous Science Series, and a lesson. Each lesson covers an organ system(s) explained with an anatomy coloring page. Do not dismiss these as mere coloring worksheets. There is a reason most college-level anatomy classes use coloring pages. There is no better way, other than dissection, to show where the major organs are in each system and how the organs work together.

With each chapter in the human anatomy portion of this course having two sections, there is more work for students to complete over the next several weeks. Therefore, these chapters are all better done on a three- to five-day schedule.

Learning Goals

- Learn the names and functions of the major organs of the nervous system.

- Learn the approximate location of the major organs of the nervous system.

- Learn the process your body uses to send information to and from your brain.

- Learn the major organs of your sensory systems.

- Learn how your sensory systems work to send and receive information.

- Learn the processes by which we hear, see, taste, and smell.

- Learn about the wavelike nature by which sound and light travel.

Extracurricular Resources

Books

It's All in Your Head: A Guide to Understanding Your Brain and Boosting Your Brain Power, Barrett, Susan

You're Smarter Than You Think: A Kid's Guide to Multiple Intelligences, Armstrong, Thomas

Understanding the Brain and the Nervous System, Sneddon, Robert

How to Really Fool Yourself: Illusions for All Your Senses, Cobb, Vicki

The Ultimate Book of Optical Illusions, Seckel, Al

Seeing, Silverstein, Alvin

Sound: More Than What You Hear, Lampton, Christopher

What Do Animals See, Hear, Smell, and Feel?, Waldrop, Victor

Online

Visit Pandia Weblinks for videos and websites recommended for this chapter:

www.pandiapress.com/weblinks-biology 2

Lesson 1

I'll Do the Thinking 'Round Here

The first lesson in Chapter 14 covers the organs in the nervous system including the brain, spinal cord, and nerve cells. These human anatomy lessons are intended for students to participate as they read, stopping when instructed to color the illustrations. The arrow students make on the illustration of the nerve cell should begin at the dendrite and end with the arrow point at the final axon.

Famous Science Series 1

Count Alessandro Volta

The subject of this chapter's first Famous Science Series is the same person featured in the nervous system lab that follows. The terms *volt* and *voltage* are named after Volta.

In 1771, biologist Luigi Galvani was in his lab cutting a frog's leg with a metal scalpel. What happened when he did? *The metal scalpel he was using touched a brass pin holding the frog's leg in place. The frog's leg twitched!*

What did he think caused it to happen? *Galvani was sure he was observing what he called animal electricity. He thought the movement was caused by electricity produced in the brain that traveled through the nerves to the muscles, making them react by moving.*

Alessandro Volta was a friend and fellow scientist. When he heard of Galvani's results, he performed the experiment. He got the same results. However, Volta did not agree with Galvani's conclusions. What did Volta think caused the frog's leg to twitch? *Volta concluded that the electricity came from contact between the metal and chemicals that were present in the frog's leg. Volta did away with the frog's leg. He used bowls of salted water instead. He made a metal arc with tin at one end of the arc and copper at the other end of the arc. Volta put the metal arc with each end in a different bowl. In this manner, he showed that the contact with the metals and the chemicals in the cells of the frog's leg were responsible for the twitching.*

Volta continued with his experiments. His experiments led him to invent something in 1800 that we still use to this day. What is it? *The battery*

Does your body produce electrical signals? Explain your answer. *Yes. The body does produce electrical signals that it sends through nerve cells. Those coupled with chemical signals that travel in the synapses between the nerves are how the brain sends and receives signals.*

Lab 1

Your Brain Is Not a Battery

It took some experimenting to find something that the lemon battery would power. It is not a strong battery. Do not try to power something other than a calculator unless you string a large series of lemons together.

Possible Answers

You are 65 percent water. Use the above discussion and the nervous system illustration you colored earlier to describe how signals are sent to and from your brain. *Using the nerves in my nervous system, electrons flow to and from my brain using water and other molecules. These electrons make electrical signals, using water as they flow back and forth.*

How would dehydration affect your brain's ability to respond to signals? Or, what would happen if the width of the wire in your lemon battery is increased or decreased? *If electrons flow through my body using water and there is not enough water, think of it like a creek that has started to dry up. In a drying-up creek, the water starts to move slowly and so do things flowing along with the water. If there is not enough water in my body, the flow of electrons starts to move slowly, which means my brain is slower responding to signals. The same thing happens when I decrease the width of the wire in a lemon battery; fewer electrons can travel through a thinner space and the signal becomes weaker. If I increase the width of the wire the signal should speed up.*

Lesson 2

Come to Your Senses

The second lesson in Chapter 14 covers the organs in the sensory system including the eye, the ear, the tongue, and the nose. The organ of touch, skin, will be covered in Chapter 15. Once again, this lesson is intended for students to participate as they read, stopping when instructed to color the illustrations.

There are quite a few parts to the eye and the ear. It is up to you how many of those you expect your student to memorize or not. Talk them through the process by which we hear, see, taste, and smell to make sure they understand how these occur. Discuss with them the function of nerves for each of these systems.

Lab 2

Seeing Sound Waves

This is a very short and easy lab. It does a good job of illustrating that sound travels as waves, though.

Possible Answers

1. Describe what you saw, heard, and felt when you struck the rubber band. *I saw the rubber band vibrate in a wavelike manner, I heard it thrum as it vibrated, and I could feel the vibrations.*

2. Using what you learned in the lemon battery lab, describe how sound waves become electrical signals your brain can interpret. *Sound waves travel through the air into my ear. They flow through my outer, middle, and inner ear, causing vibrations. Inside my ear they come in contact with water and other molecules; the vibrations are translated into electrical signals at my auditory nerve that travel to my brain. My brain then translates these signals into the sounds I hear.*

3. What did you feel when you touched the side of your throat? In words or with a picture, describe how spoken words are heard. Start with the source (the throat) and end with the sound going into an ear to the brain. *I felt vibrations when I touched the side of my throat. I make sounds in my throat and they leave through my mouth and travel to my ear, where they are translated into electrical signals.*

4. What is the process by which sound gets from its source to being heard by you? *Sound starts as vibrational waves at its source. These waves travel from their source until they get to me.*

5. Describe what you saw, heard, and felt when you struck the instrument. *It felt very similar to the rubber band. I saw the strings vibrate in a wavelike manner. I heard them thrum as they vibrated, and I could feel the vibrations.*

6. When you put your hand on the strings the sound started to quiet, why? *Because I was stopping the strings from vibrating.*

7. How do you think bats "see" in the dark? Hint: It is called *echolocation*. Can you describe what happens? *When sound waves leave their source, they go until they hit a solid or liquid surface. When they do, they "bounce" back. Bats use this to navigate and to locate food. Bats send out sound waves that travel until they hit an object. Bats use the amount of time it takes the sound waves to come back to them to figure out how far away the object is. If the object is close, the wave bounces back faster than if the object is far away. Whales and dolphins also use echolocation.*

Pandia PRESS

Famous Science Series 2

Cochlear Implants

Sticking with the theme of how people hear, the Famous Science Series for this week is about cochlear implants. There are leaps and bounds in the medical field daily, and rules about when procedures can be done change regularly too. The information and illustration are current as of 2012. I encourage you to research if the parameters and procedures regarding cochlear implants have been updated.

Describe how sound travels from its source to you so you can hear the sound? *When you hear, a sound wave travels from the source of the sound through the air to your ear. Your ear funnels the sound from your outer ear through your middle ear to your inner ear. There the sound waves stimulate your hearing nerve that sends an electrical signal to your brain. Your brain interprets the electrical signal and you hear.*

What happens if the cochlea in the ear does not work? *A person whose cochlea does not work will be deaf (unable to hear) or partially deaf.*

A cochlear implant is surgically implanted under the skin behind the ear. How does it work? *A cochlear implant has an electrode that goes from the skin into the inner ear. In the inner ear, it bypasses the cochlea and uses its own electrical signals to stimulate the hearing nerve. The nerve sends a signal to the brain and the person can hear.*

How old do you have to be in the United States before you can get a cochlear implant? *12 months old*

Show What You Know

Nervous and Sensory Systems

What are the three main parts of your brain? *cerebrum, cerebellum, medulla*
The *medulla* controls involuntary vital functions.
The *cerebrum* is responsible for complex thought.
The *cerebellum* coordinates movement.

Match the word with its definition.

Cornea — has tiny muscles that control size of eye

Pupil — white part of eye

Iris — clear layer outside of eye

Lens — behind pupil, focuses light to back of eye

Retina — light receptors

Optic Nerve — where images are formed

Sclera — black hole in the center of eye

Rods and Cones — nerve going from eye to brain

The above make up your *eye*, which is your organ of *sight*.

Label the parts of the ear.

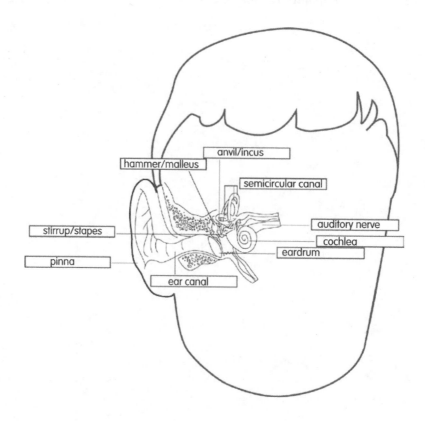

anvil/incus

hammer/malleus

semicircular canal

stirrup/stapes

auditory nerve

cochlea

eardrum

pinna

ear canal

Multiple Choice

1. The parts of your inner ear are *semicircular canals, cochlea, auditory nerve*
2. When signals travel through your nerves, the path they take is *dendrite, main body part, axon*
3. Scent molecules enter your nose, where there are *olfactory nerves* that send information about them to your brain.
4. The organs of your nervous system are *brain, spinal cord, nerve cells*
5. Sound and light travel through the air as *waves*.
6. Chemical receptors on your tongue determine if something is *sweet, sour, bitter and/or salty*.
7. The parts of your middle ear are *malleus, incus, stapes*
8. Your body's main nerve is the *spinal cord*.
9. Nerve cells send messages through your body using *electrical signals*.
10. Four organs of sense are *ear, eye, nose, tongue*

Lesson Review

Nervous and Sensory Systems

The Nervous System

The purpose of the nervous system:

- To control the other systems in your body
- To receive information about your surroundings so you can act upon that information
- To store memories and allow for thought

The primary organs in the nervous system are your brain, spinal cord, and nerves.

The brain has three parts:

1. cerebrum = complex thought
2. cerebellum = movement
3. medulla = vital function (like your heart rate)

Spinal cord = body's main nerve

Nerves send signals using electrical impulses from the dendrite → the main body nerve → axon → to the dendrite of the next nerve cell

The Sensory Systems

Use the illustrations in the book to review the sensory systems and to discuss the parts.

The purpose of the sensory systems is to send information about your surroundings to your brain.

1. Eye = sight
2. Ear = hearing
3. Taste and Smell = tongue and nose

Unit IV: Anatomy and Physiology

Chapter 15: Integumentary, Digestive, and Urinary Systems

WEEKLY SCHEDULE

Two Days

Day 1
☐ Lesson 1 & Activity
☐ Lab 1
☐ FSS 1
☐ MSLab

Day 2
☐ Lesson 2 & Activity
☐ Lab 2
☐ FSS 2
☐ Lesson Review
☐ SWYK

Three Days

Day 1
☐ Lesson 1 & Activity
☐ Lab 1
☐ FSS 1
☐ MSLab

Day 2
☐ Lesson 2 & Activity
☐ Lab 2
☐ FSS 2

Day 3
☐ Lesson Review
☐ SWYK

Five Days

Day 1
☐ Lesson 1 & Activity
☐ Lab 1

Day 2
☐ FSS 1
☐ MS Lab

Day 3
☐ Lesson 2 & Activity

Day 4
☐ Lab 2
☐ FSS 2

Day 5
☐ Lesson Review
☐ SWYK

Introduction

This chapter covers three organ systems. The primary organ of the integumentary system is your skin. The second section covers the digestive and urinary systems. These two systems process materials that come into your body. Your digestive system processes the food molecules your cells can use, and then excretes the waste they cannot. Your urinary system is responsible for maintaining the fluid balance in your body and for filtering wastes from your blood. The important concept of homeostasis as it relates to humans is introduced in this chapter.

Learning Goals

- Learn the names and functions of the major organs of the integumentary system.

- Learn the approximate location of the major organs of the integumentary, digestive, and urinary systems.

- Learn how skin helps your body maintain homeostasis.

- Learn how your body digests food.

- Learn the names and functions of the major organs of the digestive and urinary systems.

- Learn how your urinary system filters waste.

- Learn how your urinary system helps your body maintain homeostasis.

Extracurricular Resources

Books

A Journey Through the Digestive System with Max Axiom, Super Scientist, Sohn, Emily

The Digestive System: Injury, Illness and Health, Ballard, Carol

Board Game

My Food Factory, The Digestive System and Nutrition Game. This is a very good game for helping to teach the names and functions of the major organs of the digestive and urinary systems.

Online

Visit Pandia Weblinks for videos and websites recommended for this chapter:

www.pandiapress.com/weblinks-biology2

Lesson 1

Your Skin and What's Within

Once again, this is a chapter that is divided into two parts. Students learn the parts of each system as they color along while reading the lesson.

This first lesson is about the integumentary system, which is comprised of skin, hair, and nails. Caution students to be careful when they are coloring the cross-section image of skin. There are a lot of parts to it. Have them pay special attention to the blood vessels. There are two pairs, one just below the epidermis and one almost at the bottom of the dermis. For each pair, color the top blood vessel red and the bottom blood vessel dark blue.

Lab 1

Reading with Your Fingers

This lab is short, simple, and fun. The longest part to this lab is the preparation. My favorite message to write is "(student's name) loves science."

Famous Science Series 1

Louis Braille

When and where was Louis Braille born? *Coupvray, France, on January 4, 1809*

How did Braille lose his eyesight? *His father was a harness maker. At the age of three, Braille was in his father's workshop playing, when he poked himself in the eye with a sharp tool used for piercing through leather. At first, they thought the injury was not serious. Then it became infected, and antibiotics had not been invented yet. Because there were no antibiotics to kill the infection, the infection caused him to lose his sight.*

At the age of ten, Braille was awarded a scholarship to where? *The National Institute for the Blind in Paris*

What two instruments did Braille learn to play at school? *The organ and cello*

Braille invented an alphabet. Who uses it and how does it work? *The Braille alphabet uses six dots in different positions on a page, to stand for the numbers 0 through 9, and for the twenty-six letters in the alphabet. Blind people are able to use Braille to read (with their fingers) and write.*

Is Braille still in use today? *Yes. Have you ever walked into an elevator and noticed the raised dots? Did you ever feel them with your fingers? Did it feel like you were reading? That was the Braille alphabet.*

Was the Braille alphabet accepted when it was invented? *No. His students loved the alphabet he invented, but sighted officials and teachers fought against using it. It was not until 1844 when it was officially recognized. In 1878, the Braille alphabet*

was adopted by a world congress as the appropriate method of reading and writing for the blind.

Microscope Lab

The Skinny on Skin

Try to get a patch of skin with the hair still attached. This lab can be done without the stain, especially if your skin has darker tone or you have dark hair.

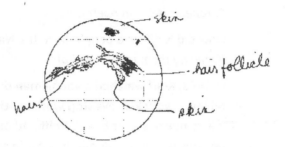

Lesson 2

Building Cells and Ridding Waste

I am talking about cell building again. Do you feel like it is a common theme? I really believe that focusing on the primarily biological reason for eating (cell building) helps children to make better nutrition choices in the long term.

Lab 2

Kidneys Clean Up

Students should plan how they are going to layer the material for their filter. When they have a plan, you should get their reasons for why they chose to layer the materials as they did. They should try different layering schemes and see which works best. If they are not happy with their results, ask them if they can think of material(s) that might do a better job. Have them try those. Try to have fun with this experiment. Some components of dirty water are easier to clean out than others. Successive steps using different materials can help with the contaminants that are harder to clean out of the water.

Famous Science Series 2

Willem Johan Kolff

The subject of the second Famous Science Series this week is a very special person. In addition to being a great scientist and a hero during the Holocaust, he was the inspiration for me in figuring out a filtering experiment for this chapter's lab.

When and where was Willem Kolff born? *Kolff was born in the Netherlands on February 14, 1911.*

Kolff was a doctor when the Nazis invaded the Netherlands. What did Kolff do when they invaded? *He was at a funeral on May 19, 1940, when the Nazis invaded. He saw the warplanes, excused himself from the funeral, and went to the city's main hospital to set up a blood bank.*

Kolff set up a special kind of bank. What was it? *He set up the first blood bank in Europe. It is still in operation.*

How did Kolff save people from the Nazi labor camps? *He saved about 800 people by hiding them in his hospital.*

In 1938, Kolff watched a young man die a painful death because his kidneys were failing. If your kidneys stop working, they do not filter waste from your body. The waste poisons your blood. Kolff and other doctors realized if they could filter the poisons out of the blood, they could save people from these painful deaths. What did Kolff do to address this problem? *Using sausage casings and technology Henry Ford had developed for his automobiles, Kolff invented the kidney dialysis machine. This machine is used by more than 200,000 Americans every year who are waiting for kidney transplants.*

Next, Kolff invented a membrane-oxygenator that is still used today. What does it do? What surgery is it still used in? *It keeps oxygen in the blood when patients are undergoing heart surgery.*

Kolff and one of his students, Dr. Robert Jarvik, invented another lifesaving device called the Jarvik 7. What is it and what did it do? *The Jarvik 7 is an artificial heart. On December 2, 1982, Barney Clark had the Jarvik 7 implanted in place of his heart. He lived 112 days with it.*

Show What You Know

Integumentary, Digestive, and Urinary Systems

Multiple Choice

1. Your largest organ is your *skin.*
2. Which part of your brain controls your digestive and urinary system? *Medulla*
3. Which nerves connect your skin to your brain? *Peripheral nerves*
4. Gastric juices in your stomach *break up food and kill bacteria.*

5. Your urinary system *filters waste from your body that your cells don't use.*
6. Skin helps maintain homeostasis by *Answers 1 and 2*
7. Your blood carries molecules to your *liver,* where they are sent to cells.
8. Melanin in the epidermis *All of the above*
9. The organ that filters your blood: *kidney*
10. Food is broken down to small molecules your body can absorb in your *small intestine.*
11. Your body makes saliva *to help break down carbohydrates.*
12. This organ is lined with muscles that mash up food into smaller bits: *stomach*
13. The color of your urine and poop comes from *digested red blood cells*
14. This is the last stop before your rectum. This is the organ that extracts the last bit of vitamins and water from your food: *large intestine*
15. The organ that fills up and then tells your brain you need to pee is the *bladder.*
16. Your skin has two main layers. They are *the epidermis and dermis.*

Questions
The three main organs of the integumentary system are:

1. skin

2. nails

3. hair

Your immune system is the system that protects you from getting infections. Your skin is sometimes called the first line of defense for your immune system. What is meant by this? *Your skin keeps pathogens, disease-causing bacteria and viruses, from getting in most places on your body. If they cannot get in, they cannot infect you.*

Lesson Review

Integumentary, Digestive, and Urinary Systems

Integumentary System
The purpose:
- to protect the inside of your body
- to send sensory information about touch to your brain
- to help maintain homeostasis

The primary organs are your skin, nails, and hair.

Homeostasis is the maintenance of an organism's internal conditions within a normal range.

Just like plants, one condition people need to maintain homeostasis for is the concentration of water in their body. Using the integumentary system and urinary system, your body maintains a healthy water level.

Water is important because the chemical reactions that drive your bodily functions take place in WATER! Water is the main component in the fluids that carry cell-building and energy-making molecules throughout your body. Water is the main component in the fluids that carry waste material out of your body.

Digestive System

Why do you eat? To build cells. The digestive process is how your body breaks down the material it needs to make cells.

The purpose: to take the food you eat and mash it down into molecules your cells can transport across their cell membranes

The primary organs are your mouth, salivary gland, esophagus, stomach, small intestine, liver, pancreas, gallbladder, large intestine, rectum, and anus.

Urinary System

The purpose:

- to remove waste
- to help maintain homeostasis by balancing the amount of fluid in your body

The primary organs are your kidneys, ureters, bladders, and urethra.

Unit IV: Anatomy and Physiology
Chapter 16: Endocrine and Reproductive Systems

WEEKLY SCHEDULE

Two Days

Day 1
- ☐ Lesson 1 & Activity
- ☐ Lab 1
- ☐ FSS 1

Day 2
- ☐ Lesson 2 & Activity
- ☐ Lab 2
- ☐ FSS 2
- ☐ Lesson Review
- ☐ SWYK

Three Days

Day 1
- ☐ Lesson 1 & Activity
- ☐ Lab 1
- ☐ FSS 1

Day 2
- ☐ Lesson 2 & Activity
- ☐ Lab 2
- ☐ FSS 2

Day 3
- ☐ Lesson Review
- ☐ SWYK

Five Days

Day 1
- ☐ Lesson 1 & Activity

Day 2
- ☐ Lab 1
- ☐ FSS 1

Day 3
- ☐ Lesson 2 & Activity

Day 4
- ☐ Lab 2
- ☐ FSS 2

Day 5
- ☐ Lesson Review
- ☐ SWYK

Introduction

Two closely related systems are covered in Chapter 16: endocrine and reproductive. Both systems are different for males and females. I realize reproduction can be a sensitive subject, but this is biology. Sexuality, intercourse, sexually transmitted diseases, and contraception are not discussed in this course.

Learning Goals

- Learn the names and functions of the major organs of the endocrine system.

- Learn the approximate location of the major organs of the endocrine system.

- Learn about hormones and how they send chemical messages through your body to regulate metabolic activity and maintain homeostasis.

- Learn the concept and process of homeostatic feedback mechanisms.

- Experiment and observe your body maintaining homeostasis.

- Learn the names and functions of the major organs of the reproductive system.

- Learn the approximate location of the major organs of the reproductive system.

- Learn about the menstrual cycle.

- Learn how a boy's and girl's body changes as they go through puberty.

- Learn the process humans go through inside their mother as they develop from a zygote to a child.

- Learn about the birth process.

Extracurricular Resources

Books

The Reproductive System, How Living Creatures Multiply, Silverstein, Alvin, and Silverstein, Virginia

The Endocrine and Reproductive Systems, Kim, Melissa L.

Ending the Food Fight, Guide Your Child to a Healthy Weight in a Fast Food/Fake Food World, Ludwig, David

Online

Visit Pandia Weblinks for videos and websites recommended for this chapter:

www.pandiapress.com/weblinks-biology2

Lesson 1

The Balancing Act

There is a male and female illustration of the endocrine system to color as students read the lesson. You could have boys color the boy and girls color the girl, but do have them identify the location of the ovaries or testes for the opposite sex.

Lab 1

Let's Maintain Homeostasis

After you have demonstrated the feedback mechanism using your thermostat, ask students to repeat back to you what a homeostatic feedback mechanism is.

1. *Temperatures will vary.*

   How does this demonstrate a feedback mechanism? *The thermostat turns on and off to maintain the temperature at which the thermostat is set.*

2. What happened to the pupils of the person when you shined the light in their eyes? *The pupils were bigger before I shined a light in them.* Explain how that is an example of maintaining homeostasis. *Your pupils control the amount of light coming into your eyes. They get bigger to let more light in, and they become smaller to let less light in. This getting bigger and smaller is how your eyes keep the amount of light coming in within a certain range.*

3. What happened when you balanced on one foot? *I swayed to maintain balance.* What happened when you swayed? *I was able to maintain balance.* How is that an example of maintaining homeostasis? *My body is compensating for being on one foot to keep me upright so I can stay standing, which was my normal condition.*

4. *Pulse rates will vary.*

   What happened to your pulse rate when you were moving quickly? What happened to your pulse rate while you rested? *When I moved quickly, my pulse rate increased. When I rested, my pulse rate decreased.* How does your pulse rate increasing when you are active and decreasing when you are at rest demonstrate a feedback mechanism? *When you need more oxygen, your body responds by pumping blood that is carrying oxygen through it at a faster rate so it gets more oxygen; this makes your pulse rate increase. When you are resting, your body needs less oxygen, so everything slows down. That is what feedback mechanisms do; they respond to specific needs or signals, keeping things within specific parameters.*

Famous Science Series 1

Body Weight and Homeostasis

If you find the topic of this Famous Science Series as interesting as I do, I recommend you read *Ending the Food Fight, Guide Your Child to a Healthy Weight in a Fast Food/Fake Food World* by David Ludwig.

One of the most important jobs your endocrine system has is maintaining homeostasis. What things does your endocrine system monitor for homeostasis? *body temperature, balance of water and other chemicals, body weight*

Why is staying at a weight part of homeostasis? *Staying within a certain body weight range means you are in a stable environment.*

What is the job of the hypothalamus? *Your hypothalamus controls your metabolism and body weight. It monitors the nutritional needs of your body, making sure your body has all the cell-building, energy-making molecules it needs. Your hypothalamus works to maintain a stable body weight by directing hormones to take some food for immediate use and some food as storage for later. The stored food translates to body weight.*

How can it become a problem that maintaining your weight is part of homeostasis? *The problem can be if your body gets used to a weight that is high and unhealthy. It is hard to lose weight if you have been at a weight for a while, because that becomes the point of homeostasis. You might have noticed, if you have ever quickly gained weight, you do not feel good at your new increased weight, but you will if you stay there too long.*

Lesson 2

To Be

There is a male and female illustration of the reproductive system to color as students read the lesson. This time I recommend you have students color both sexes because the organs are completely different.

Lab 2

Your Story

In Appendix C of the student Workbook, there is a five-paragraph essay worksheet if you choose to have your student write a formal essay of his birth.

This lab is a chance for your student or child to learn his unique birth story. Read over the text from the chapter and try to use as much vocabulary and "science words" as possible when you tell his story. This way your child will hear the vocabulary in a context that should help him or her remember and relate to it. If possible have your child listen to birth stories other than his own. Focus on how these labors and deliveries were different from his own. If you have a film or photos of your child's birth and are comfortable with it, this would be a good time to watch it together.

If your student doesn't have access to his own birth story, have him learn the birth stories of other family members or close friends.

Famous Science Series 2

Famous Doubling: Twins

The topic in this Famous Science Series is how the two different types of twins occur. Monozygotic twins (also called identical), form from one zygote = *mono* zygote. Dizygotic twins (also called fraternal) form from two zygotes = *di* zygotes.

How many zygotes make monozygotic twins? *One*

Are monozygotic twins identical or fraternal? *Identical*

What is the process that creates monozygotic twins? *Monozygotic twins result when one sperm fertilizes one egg. Shortly after fertilization, two embryos are formed through the process of mitosis.*

How many zygotes make dizygotic twins? *Two*

Are dizygotic twins identical or fraternal? *Fraternal*

What is the process that creates dizygotic twins? *Dizygotic twins results when two eggs are fertilized by two sperm.*

Do dizygotic twins have the same DNA? *No. Dizygotic twins share the same birthday, but they are as close genetically as any offspring from the same parents.*

**Show What
You Know**

Endocrine and Reproductive Systems

Multiple Choice

1. The purpose of the endocrine system is to *All of the above*

2. The birth process has three stages. The order they follow is *labor, delivery of baby, delivery of placenta.*

3. Menstruation occurs when *the female gamete is not fertilized.*

4. Your endocrine system is made of *endocrine glands*

5. An example of a homeostatic feedback mechanism is *insulin decreasing glucose in your blood after you have eaten a big meal and glucagon increasing glucose in your blood when glucose gets low*

6. What is the purpose of the umbilical cord? *It transports oxygen and nutrients from the mother to the fetus, and wastes from the fetus to the mother.*

7. Fertilization occurs in the *fallopian tubes.*

8. The chemical messengers that send information back and forth between organs are *hormones*

9. The testes are outside a male's body *so the sperm does not get too hot.*

10. At eight weeks the embryo is called a *fetus.*

11. What is the purpose of the male reproductive system? *To make and deliver the male gamete*

12. Females make eggs: *They are born with them already made.*

13. An example of homeostasis is *sweating when it is hot.*

14. The fetus develops in the *uterus.*

15. What is the purpose of the female reproductive system? *All of the above*

16. Puberty begins when the endocrine system begins making more of the hormone *testosterone* in boys and the hormone *estrogen* in girls.

Lesson Review

Endocrine and Reproductive Systems

Endocrine System

The purpose: to produce and deliver hormones. Hormones regulate metabolic activity and are important to your body's ability to maintain homeostasis.

The primary organs are your hypothalamus, endocrine glands, thymus, pancreas, ovaries (in girls), and testes (in boys).

Homeostasis = maintain internal condition within a normal range
Homeostatic feedback mechanism = the mechanism your body uses to keep internal conditions within the normal range
Hormones = the chemical messengers your body uses to maintain homeostasis, to keep conditions within the normal range

Your blood carries hormones through your body to cells with receptors on them. When receptors come in contact with hormones, your body responds.

Hormones control bodily functions such as:
- cell division
- the rate of digestion
- blood pressure
- hunger
- production of white blood cells
- the development of male and female bodies during puberty
- menstruation
- the amount of glucose in your blood

Reproductive System

The purpose: to make us

The primary organs for boys are the testes, scrotum, urethra, and penis.
The primary organs for girls are the ovaries, fallopian tubes, uterus, cervix, and vagina.

Making More

1. Meiosis → sexually reproducing organs make gametes with one set of chromosomes in each gamete
2. Fertilization → in the female's fallopian tube fertilization occurs, the male and female gametes fuse → the fused gametes make a zygote with two sets of chromosomes, one from the mother and one from the father
3. Mitosis → the zygote divides → then those cells divide → and so on; all these cells are genetically identical but specialize, forming different tissue types and organs
4. Inside the mother the growing child attaches to her through the umbilical cord. With the nutrients delivered through the cord, the fetus gets the cell-building molecules it needs to continue to grow more cells.
5. After about nine months, the organs of the fetus are developed enough to be born.
6. The mother goes into labor and a baby is born.

Unit IV: Anatomy and Physiology
Chapter 17: Circulatory and Respiratory Systems

WEEKLY SCHEDULE

Three Days

Day 1
- ☐ Lesson 1 & Activity
- ☐ MSLab
- ☐ Lab 1
- ☐ FSS 1

Day 2
- ☐ Short Story
- ☐ Lesson 2 & Activity
- ☐ Lab 2
- ☐ FSS 2

Day 3
- ☐ Lesson Review
- ☐ SWYK

Five Days

Day 1
- ☐ Lesson 1 & Activity
- ☐ MSLab

Day 2
- ☐ Lab 1
- ☐ FSS 1

Day 3
- ☐ Short Story
- ☐ Lesson 2 & Activity

Day 4
- ☐ Lab 2
- ☐ FSS 2

Day 5
- ☐ Lesson Review
- ☐ SWYK

Introduction

The organ systems covered in Chapter 17 are the circulatory and respiratory systems. These two systems work together to deliver oxygen throughout your body and carbon dioxide out of it. In addition, the circulatory system transports nutrients and water throughout your body and works with your digestive and urinary systems to get rid of waste. This is a long chapter even for this unit. For that reason, only three- and five-day schedules are offered.

Learning Goals

- Learn the names, functions and locations of the major organs of the circulatory system.

- Learn the names and functions of the four main components of blood.

- Learn how the heart and lungs get oxygen into blood and carbon dioxide out of it.

- Investigate how doctors amplify the sound of your heartbeat.

- Learn the names, functions, and locations of the major organs of the respiratory system.

- Learn how you make sound.

- Learn about lung capacity.

Extracurricular Resources

Books

Understanding the Human Body–The Circulatory System, Walker, Pam and Wood, Elaine

The Circulatory System (The Human Body: How It Works), Whittemore, Susan

101 Questions About Blood and Circulation: With Answers Straight From the Heart, Brynie, Faith

The Respiratory System: How Living Creatures Breathe, Silverstein, Alvin

Understanding the Human Body–The Respiratory System, Walker, Pam and Wood, Elaine

Sacred Mountain: Everest, Taylor-Butler, Christine

Living in the Himalayas, Spilsbury, Louise

An excellent book about a failed Everest expedition. It is written for adults, but very accessible for students: *Into Thin Air: A Personal Account of the Mt. Everest Disaster*, Krakauer, Jon

Online

Visit Pandia Weblinks for videos and websites recommended for this chapter:

www.pandiapress.com/weblinks-biology 2

Lesson 1

Going Round in Circles

There are three distinct parts to this section on the circulatory system:

1. The names and functions of the organs of the circulatory system.
2. The cycle and path that blood flows, in order to become oxygenated and to get rid of waste.
3. The names and functions of the four main components of blood.

When you cut yourself, the blood you see is always red. Why? *This is because cuts are open to the air. About 21 percent of the molecules in air are oxygen molecules. When a cut occurs, the blood immediately encounters oxygen, and any red blood cells that don't have oxygen in them take it up right away.*

Microscope Lab

OUCH!

If your students are like mine, prepare to stab your finger. Every time I have done this experiment, the only person stabbed was me. The good news is you do not need much blood. You will be able to see the components of blood. It really is worth it.

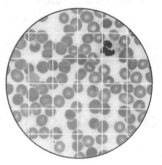

Lab 1

The Beating Heart

The diameter of the tubing can vary depending on whether you fit it in the outside or the inside of the bottle. Use duct tape to connect the bottle to the tubing. No other tape I tried worked as well.

We listened to heartbeats, lungs, and stomachs of people, cats, guinea pigs, and dogs. Despite long hair, the heartbeat that was easiest to hear was that of a border collie.

Answers to the questions on the lab sheet will vary.

Famous Science Series 1

Rene-Theophile-Hyacinthe Laënnec

The Famous Science Series and the lab for the circulatory system were paired because of their relationship to each other. Rene-Theophile-Hyacinthe Laënnec invented the stethoscope.

What did Rene-Theophile-Hyacinthe Laënnec invent? When did he invent it? *The stethoscope, in 1816*

Laënnec's mother died when he was five years old from what disease? What does this disease do to people who have it? Was this disease serious and has a cure been found? *Laënnec's mother died from tuberculosis. Tuberculosis is a disease that if left untreated causes extensive damage to the lungs. Before antibiotics were discovered, tuberculosis was the leading cause of death in many countries. It had a 50 percent mortality rate. 90 percent of people with tuberculosis who receive treatment are cured using antibiotics for 6 to 12 months. Antibiotic-resistant tuberculosis is on the rise, though. Even now, tuberculosis kills 2 to 3 million people every year.*

How did Laënnec get the idea for his invention? *He was taking a walk when he observed two children playing with a piece of solid wood and a pin. One child put his head to one end of the wood, and the other child scratched the pin on the other side of the wood. The scratching of the pin was amplified along the wood to the child with his ear to it.*

Later the same year Laënnec had a patient who looked to be suffering from heart disease. The standard treatment at that time was for the doctor to put their ear to the patient's chest. This patient was a young, plump female and Laënnec was uncomfortable putting his head to her chest. At that moment, he remembered the children with the wood and pin. He rolled up a sheet of paper and put that to her chest. He was surprised by how clearly he could hear her heart.

How did he name his invention? *He named the stethoscope from the Greek words* stethos, *which means "chest," and* scope, *which means "observer."*

Short Story

"The Tell-Tale Heart"

I like to pair literature and science. Who better than Edgar Allan Poe to pair with the circulatory system? "The Tell-Tale Hart" is available free online.

Possible Answer

After reading the story, explain scientifically, why it was or was not possible for the old man's heart to be still beating. *Organs function as a part of an organ system. When an organ is removed from its organ system, the organ no longer works. In addition, a person's brain is needed to make the person's heart beat. The old man no longer has his brain connected to his heart.*

Read

Lesson 2

Breathe In, Breathe Out

What organisms benefit from the carbon dioxide you breathe out? Plants. The respiratory section is a good place for a review of photosynthesis and cellular respiration.

Do you remember what the process is called when molecules go from an area of high concentration to an area of low concentration? *diffusion*

Explore

Lab 2

Lung Capacity

This lab requires at least two people to perform it. In addition, I recommend teacher supervision. In most cases, different people have different lung capacity. It is interesting to compare lung-capacity results for different people and discuss what parameters lead to the differences in capacity. For example, bigger people have a larger lung capacity. Athletes and more active children have a larger lung capacity than their peers who are less active. People with respiratory ailments and those who smoke generally have a smaller lung capacity.

The math this week will be challenging for some students who have never been exposed to geometry problems dealing with circles. They are just formulas, though. The use of a calculator is highly recommended. Even if you have to do the problems with your student, these types of problems are valuable to expose students to. The type of analysis done in this experiment is one of the reasons for learning the formulas in the first place.

Included in the lab is a math review of definitions and calculations of circumference, radius, diameter, and volume. Radius can also be calculated by dividing the diameter by 2, but it is hard to measure the diameter of a balloon shape. Therefore students calculate the radius using the circumference of the balloon, a much easier aspect to measure on a balloon.

Most of the answers on the lab sheet will vary considerably due to difference in measurements and calculations.

Possible Answers

Were you surprised by the expiratory reserve in your lungs? Why or why not? *Answers will vary*

Why did you take 5 measurements? *I took 5 measurements so that I would have an average. Good science is not about getting 1 good number. It is about repeating an experiment and getting several measurements to get an overall answer, an average, which gives you a more accurate understanding of what is happening.*

How do you think VLC would change for someone who smoked? Why? *VLC would decrease for someone who smoked because their lungs would have tar in them, taking up space.*

Math This Week

In both labs, there is counting and multiplying. In the respiratory system lab, students calculate circumference, radius, and the volume of a sphere.

Pandia PRESS

How do you think VLC would change for someone who was a marathon runner? Why? *VLC would increase from stretching and use, as a marathon runner needed more oxygen and therefore breathed harder and more often.*

Famous Science Series 2

Sherpas

This Famous Science Series uses the term *concentration*. Concentration is a measure of the amount of particles in a given volume. If this is a new term for your student, here is an easy way to demonstrate the concept: Make a drink that requires the addition of sugar, e.g. iced tea or lemonade. When you make the drink, make separate batches, with varying amounts of sugar. Monitor and discuss the different concentrations of sweetener. Taste the different solutions. Then discuss how the different concentrations of sweetener affect the taste. Discuss how the concentration of sugar in a liquid is similar to the concentration of different molecules in the air.

Where do Sherpas live? *In Nepal, high in the Himalaya Mountains at altitudes between 3,000 and 4,300 m (10,000 to 14,000 ft)*

Where does the name *Sherpa* come from? *Sherpa means "easterner," because they come from the Kham in Eastern Tibet.*

What mountain do Sherpas help people climb, and how do they help them climb it? How high is this mountain? *For decades, the Sherpa people have been the preferred guides and porters for people climbing Mount Everest. At 8,850 meters tall and growing (it grows 4 to 10 cm a year because of geological forces), it is the tallest mountain on Earth.*

How much less is the concentration of oxygen molecules at the top of Mount Everest? *The concentration of oxygen molecules is 33 percent less than at sea level. This translates to 33 percent fewer oxygen molecules in every breath taken at the top of Mount Everest.*

Why is this a problem? *The problem is your body still needs the same amount of oxygen it did at sea level.*

Sherpas have a larger lung capacity than people who live at lower elevations. How does this help them deal with lower levels of oxygen? *Sherpas are made of cells just like everybody else, and those cells need the same amount of oxygen as everybody else's. When a group of Sherpas were tested to look at their lung capacity, it was discovered that they have a larger lung capacity than average. Scientists think this larger lung capacity is an adaptive response to living a very active lifestyle in an area where the concentration of oxygen is low for humans. A larger lung capacity means that Sherpas get more air and therefore more oxygen molecules into their lungs with every breath they take.*

Show What You Know

Circulatory and Respiratory Systems

Multiple Choice

1. Your arteries *carry blood away from your heart to the rest of your body.*
2. Your capillaries *are one cell thick, which allows transport through them.*
3. Your veins *carry blood to your heart from your lungs.*
4. Mucus *traps germs so they don't get into your lungs.*
5. Your alveoli and capillaries work together to *transport carbon dioxide into your lungs and oxygen into your blood.*
6. The muscle under your lungs that helps you breathe is your *diaphragm.*
7. The protein molecule that carries oxygen in your red blood cells is *hemoglobin.*
8. *Platelets* rush to the site of a cut and produce *fibrin*, which forms a protective covering over the cut.
9. The tubes that go into your lungs are called *bronchi.*
10. Plasma in your blood carries *All of the above*
11. Your heart has two sides that pump blood: *the left side pumps blood out and the right side pumps blood in*
12. The part of your brain that controls your heartbeat and your breathing is your *medulla.*
13. Vocal cords are two folds of tissue that stretch across this organ, allowing you to talk. Your vocal cords are in your *larynx.*
14. When you eat and drink, this closes over the entrance of your trachea so no food or drink gets into your lungs: *epiglottis*
15. Your skin and white blood cells fight infections together by: *your skin prevents most germs from getting in, but when they do, white blood cells destroy them*

Lesson Review

Circulatory and Respiratory Systems

Circulatory System

The Purpose: To carry blood throughout your body. Blood is how your body transports food and other necessary molecules to your cells. Blood is how your body transports waste away from your cells to places where your body can eliminate the waste molecules.

The primary organs are your heart, arteries, veins, and capillaries.

The job of your circulatory system is to transport material throughout your body. It uses arteries, veins, and capillaries to do this.

Arteries–take blood AWAY from heart

Veins–take blood TO heart

Branching off arteries and veins are **capillaries** one-cell thick. Material moves through the thin walls of capillaries.

Parts of Blood

Red blood cells–contains the protein molecule hemoglobin, which transports oxygen and carbon dioxide

White blood cells–attack bacteria and viruses keeping them from infecting you

Platelets–produce fibrin, which form scabs and clots to heal cuts

Plasma–the fluid red blood cells, white blood cells, and platelets are in. Plasma transports food, waste molecules, and hormones

Go back over the Blood Flow Diagram to review the path blood flows through the body.

Respiratory System

The Purpose:

- to deliver oxygen from the air to your blood
- to take carbon dioxide from your blood and expel it into the air
- to make sounds

The primary organs are your nose and mouth, larynx, trachea, bronchi, and lungs.

Your **larynx** has two folds of skin across it called your **vocal cords.**

When air goes past tightened muscles in your vocal cords sound is made.

Use the Respiratory System Diagram to follow the path and the organs involved in respiration.

Unit IV: Anatomy and Physiology

Chapter 18: Skeletal and Muscular Systems

WEEKLY SCHEDULE

Two Days

Day 1
- ❑ Lesson 1 & Activity
- ❑ Lab
- ❑ Lesson 2 & Activity

Day 2
- ❑ Dissection & MSLab
- ❑ FSS
- ❑ Lesson Review
- ❑ SWYK

Three Days

Day 1
- ❑ Lesson 1 & Activity
- ❑ Lab

Day 2
- ❑ Lesson 2 & Activity
- ❑ Dissection & MSLab

Day 3
- ❑ FSS
- ❑ Lesson Review
- ❑ SWYK

Five Days

Day 1
- ❑ Lesson 1 & Activity
- ❑ Lab

Day 2
- ❑ Lesson 2 & Activity

Day 3
- ❑ Dissection & MSLab

Day 4
- ❑ FSS
- ❑ Lesson Review

Day 5
- ❑ SWYK

Introduction

The organ systems covered in Chapter 18 are the muscular and skeletal systems. These two systems are so interdependent that together they are called the musculoskeletal system.

Learning Goals

- Learn the names and purposes of the major organs of the skeletal system.

- Identify by name many of the bones in the skeletal system.

- Learn that bones are living tissue.

- Learn the structure of bones.

- Learn how bones make blood cells.

- Learn about and identify the six different kinds of movable joints in the human body.

- Learn the purpose of the muscular system.

- Identify by name many of the muscles in the muscular system.

- Learn the names and functions of the three types of muscle tissue.

- Learn how the skeletal and muscular systems coordinate. forming the musculoskeletal system that is responsible for coordinated movement.

Extracurricular Resources

Books

Skeletal System (Invisible World), Avnau, Edward

The Illustrated Guide To the Human Body: Skeletal and Muscular System, Bender, Lionel

How Our Muscles Work (Invisible World), Avila, Victoria

The Skeletal System Frameworks for Life, Silverstein, Alvin, and Silverstein, Virginia

Online

Visit Pandia Weblinks for videos and websites recommended for this chapter:

www.pandiapress.com/weblinks-biology2

Read

Lesson 1

A Structured Life

The names of bones are included as a matter of interest and information. It is up to the instructor whether students need to memorize the names of the bones.

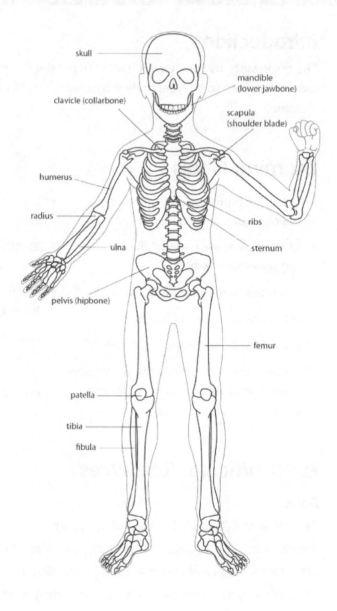

skull

mandible
(lower jawbone)

clavicle (collarbone)

scapula
(shoulder blade)

humerus

radius

ribs

ulna

sternum

pelvis (hipbone)

femur

patella

tibia

fibula

Explore

Lab

The Joint Detective Game

Before starting this lab make sure your students have looked over the Six Types of Joints and understand what type of movement can be done with each joint type. This lab can be long or short depending on how much time you allot and how much fun your students have finding different types of joints.

Pandia PRESS

Possible Answers

	Ball-and-socket joint	Hinge joint	Pivot joint	Saddle joint	Gliding joint	Ellipsoid joint
BODY	Shoulder Hip	Knee Elbow Fingers Toes Jaw	Top of neck Spine	Thumb (The base of the thumb is the only saddle joint on the human body)	Some of the joints in the foot Some of the joints of the wrist (carpal bones)	Finger knuckles Wrist (radio-carpal joint)
HOME	PS2 game controller Knob Camera tripod	Front door Toy dump truck Folding chair Switch Books Scissors Gun trigger Hole punch Cabinet doors	Doorknob	Adjustable TV stand Rider on a saddle (where the name comes from)	Gun Telescope Telescopic antenna Casters on a rolling chair Trombone slide	

Lesson 2

I Can't Sit Still

The names of muscles are included as a matter of interest and information. It is up to the instructor whether students need to memorize the names of the muscles.

Dissection and Microscope Lab

A Chicken Wing Thing

Students should take their time with this lab. If they are careful and methodical, they will see a clear demonstration of how the musculoskeletal system works to coordinate movement. This lab also shows the organs of the muscular and skeletal systems. Make sure the chicken wings are fresh.

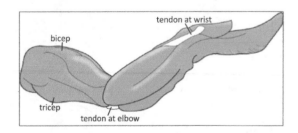

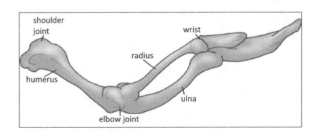

There will be ligaments and cartilage at the elbow joint. The ligaments are between the bones holding them together, and the cartilage is the slippery material between the bones that helps the bones glide against each other.

Possible Answers

1. Describe what happened in your muscle and skeletal systems when you raised the book. *My muscles and skeleton worked together to lift the book. The muscles pushed and pulled; the ligaments kept the elbows together; the tendons connected the bones and muscles so they could push the book up.*

2. What do you see at the cut-away shoulder joint? Do you think the chicken wing came from the right or left side of the body? Why? *Answers may vary, possibilities are ligaments, cartilage, muscle, tendons, skin, bone*

3. Describe the chicken skin. The skin is a part of what organ system? *Integumentary.* When you cut away the skin, there was tissue connecting the skin to the muscle underneath it. What is this type of tissue called? *Connective tissue.* What is the name of the layer of skin on the outside of the chicken wing? *Epidermis.* What is the name of the skin on the inside that is connected to the muscle? *Dermis*

4. Which muscle in the upper arm of the chicken wing is a flexor and which is an extensor? *The biceps is a flexor and the triceps is a an extensor*

5. Why are tendons attached to the ends of the muscles? *So that when the muscles move, the bone does too.* Tendons are made from one type of tissue. What type do you think that is? Why? *Connective tissue, because tendons connect muscles and bones*

6. Why is there a blood vessel running down the leg of a chicken? Blood vessels are a part of what organ system? *Legs need oxygen and other nutrients for energy and to get rid of carbon dioxide and waste products; legs use blood to do these things and blood vessels to carry the blood. Blood vessels are part of the circulatory system.*

7. Describe how movement was affected when one muscle in a pair was cut. *It was stopped.*

8. What kind of joint is the elbow joint? How does this compare with your elbow joint? *Hinge joint, it is the same as my elbow joint*

9. How do the three bones at the elbow joint fit together? *Answers may vary*

10. Describe with words, a drawing with labels, or a labeled photograph, the inside of the bone. What is the name of the soft, red material inside the bone? What important function does this material provide? *Marrow or spongy bone; it is where blood cells are made*

11. How does form fit function? Now that you have looked at a blood vessel, why do you think red blood cells have the shape they do? *Answers will vary, but it is amazing how the cells form different shapes depending on their specialization.*

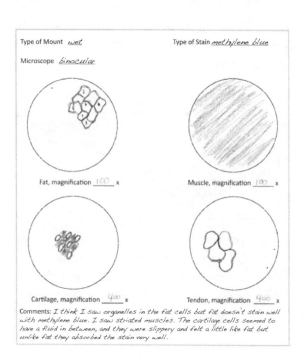

Type of Mount *wet*   Type of Stain *methylene blue*

Microscope *binocular*

Fat, magnification ___100___ x   Muscle, magnification ___100___ x

Cartilage, magnification ___400___ x   Tendon, magnification ___400___ x

Comments: *I think I saw organelles in the fat cells but fat doesn't stain well with methylene blue. I saw striated muscles. The cartilage cells seemed to have a fluid in between, and they were slippery and felt a little like fat but unlike fat they absorbed the stain very well.*

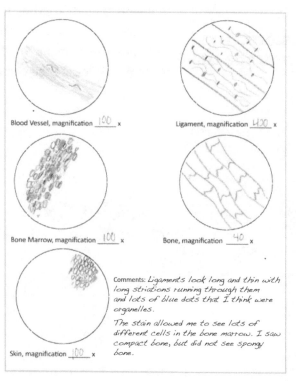

Blood Vessel, magnification ___100___ x   Ligament, magnification ___400___ x

Bone Marrow, magnification ___100___ x   Bone, magnification ___40___ x

Skin, magnification ___100___ x

Comments: *Ligaments look long and thin with long striations running through them and lots of blue dots that I think were organelles.*

The stain allowed me to see lots of different cells in the bone marrow. I saw compact bone, but did not see spongy bone.

**Famous Science
Series**

Perfecting Prosthetics

My paternal grandfather had two prosthetic legs. He was fitted for them in the 1950s. As kids, we loved it. He would take his legs off and we had a grown-up who was our height. If they were uncomfortable we didn't know it. He never complained. His gait when he walked was stilted, though. He was a pilot, and lost both legs in a plane accident. He continued to fly planes using his prosthetic legs. Scientists have come a long way in the field of prosthetics since my grandfather was fitted for his prosthetic legs in the 1950s.

What does the word *prosthetic* mean? *artificial limbs*

How long have people been using them? What were they made from in ancient times? *People have been using prosthetics for thousands of years. People lose limbs for many reasons. Thousands of years ago, before man-made plastic was invented, people made prosthetics from metal or wood.*

What four organ systems have to be coordinated to make a limb move? *The nervous, integumentary, skeletal, and muscular systems*

**Show What
You Know**

Skeletal and Muscular Systems

Multiple Choice

1. How does shivering help your body maintain homeostasis? *When you are cold, your muscles generate heat by shivering.*
2. Skeletal muscles are *voluntary muscles.*
3. Involuntary muscles move *without you thinking about them moving.*
4. Voluntary muscles move *when your brain tells them to move.*
5. When red blood cells are made, they head to your *heart.*
6. What organ is made of cardiac muscles? *Heart*
7. Your muscles are attached to bones by *tendons.*
8. Your bones are attached to each other by *ligaments.*
9. The slippery material at joints where two bones meet is called *cartilage.*
10. Smooth muscles *line blood vessels.*
11. The process where cartilage parts turn into bone as you age is called *ossification.*
12. The outer covering on bones is called *compact bone.*
13. The webbed part of bone that allows compression and absorbs force is called *spongy bone.*
14. The point where two bones meet is a *joint.*
15. Blood cells are made in *spongy bone.*

Lesson Review

Skeletal and Muscular Systems

Skeletal System

The Purpose
- to protect your internal organs
- to give you support and shape
- to work with your muscular system so you can move

The primary organs are your bones, ligaments, cartilage, and tendons.

Bones are living tissue.

There are two types of bone:

1. **Spongy bone** = webbed to allow for compression and withstand force = where blood cells are made
2. **Compact bone** = dense, protective outer layer of bone

ligaments = connect bones to bones at joints
cartilage = in between bones at joints to help bones glide against each other
tendons = connect bones to muscles
joints = the site where two bones meet

Muscular System

The Purpose
- to work with your skeletal system so you can move
- to help maintain homeostasis

The primary organs are your muscles.

Involuntary muscles move without you thinking about them moving.
Voluntary muscles move when you think about them moving.

There are three types of muscle tissue in your body:

1. **Smooth muscles** = involuntary. They control breathing, digestion, and blood pressure. They line the inside of many parts of your body so that things run smoothly through your body.
2. **Cardiac muscles** = only in your heart, involuntary muscles
3. **Skeletal muscles** = voluntary muscles. They are striated muscles (striated = banded). These are the muscles that work with your bones so you move; they are attached to your skin and your bones.

Unit IV: Anatomy and Physiology
Chapter 19: Immune and Lymphatic Systems

Introduction

The organ systems covered in Chapter 19 are the immune and lymphatic systems. These two systems function together to keep you healthy. The second half of the week examines how the organ systems of the human body work together to make you, the organism.

This is the last chapter in Unit IV. There is a Unit IV exam that covers the material found in Chapters 11 through 19, in the appendix of the student Workbook. The answer key is found at the end of this chapter.

Preparation for the next chapter: Begin to collect dead winged insects for the microscope lab in Chapter 20.

Learning Goals

- Learn the names and functions of the immune and lymphatic systems.

- Learn how your body protects itself from pathogens.

- Learn the purpose of vaccines and the benefits of immunity.

- Learn how white blood cells fight and recognize pathogens.

- Learn what exponential growth means and how this can lead to a situation where infections can make you sick.

- Investigate how organ systems work together so that human organisms function.

Extracurricular Resources

Books

Immune System: Your Magic Doctor, Garvy, Helen

Dr. Jenner and the Speckled Monster: The Discovery of the Smallpox Vaccine, Marrin, Albert

Smallpox in the New World (Epidemic!), True Peters, Stephanie

Online

Visit Pandia Weblinks for videos and websites recommended for this chapter:

www.pandiapress.com/weblinks-biology2

The Warrior Systems

The administration of vaccines can be a controversial subject. In this chapter, I do not deal with the controversy. I explain the purpose and importance of vaccines without dealing with any of the potential side effects or rare risks.

Bacteria Out of Control

Lab 1

This is a very short lab demonstrating how quickly a pathogen can go from one individual to many. This might be the first time your student hears about things growing exponentially. Other examples of exponential growth you could discuss are human population and compound interest.

Answers

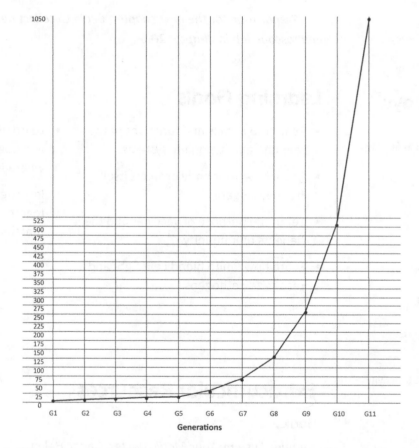

Math This Week

1. Exponential growth
2. Multiplication
3. Plotting points on a chart and drawing the exponential growth curve through those points

1. If it takes bacteria 15 minutes to divide, how long would it take for 11 generations to go from G1 to G11? *Going from G1 to G11 the bacteria divides 10 times. 10 x 15 = 150 minutes, which is 2½ hours.*

2. One bacterium can become 1 million bacteria in seven hours. How does that make it hard for your immune system to respond? *You have to make enough white blood cells to kill all those bacteria. That can overwhelm your system. That is a lot of white blood cells.*

3. This type of growth is called *exponential growth*. Look up *exponential growth* in the dictionary and write the definition here. *Growth where the rate of growth becomes ever more rapid in proportion to the growing total number.*

Famous Science Series

Smallpox: What Columbus Brought

The death toll from some diseases can be staggering. Smallpox was one example. In David McCullough's book *1776*, he relates that colonists feared the British abandoning Boston and leaving behind people infected with smallpox, who would infect the Continental Army. Because of this concern, the first group of Continental soldiers General George Washington sent into Boston were all men who had contracted and survived smallpox and were therefore immune to the disease.

I think this is a very interesting area of history. For a great adult book on the subject of plague I recommend: *Justinian's Flea: Plague, Empire, and the Birth of Europe* by William Rosen. It gives you a real idea of how random and capricious killer diseases have been over the course of history, where a fluke of an individual's immune system, not merit, determines who survives and who does not.

There have been a few types of pathogens that over the course of human history have caused much death and suffering. The smallpox virus is one of those. What happens to people infected with it? *People infected with it get small bumps, about the size of grains of rice, filled with pus all over their body. They also get a fever; and they can die.*

When and where did smallpox originate? *In Africa over 12,000 years ago. The smallpox virus has been found on Egyptian mummies that are 3,000 years old.*

Was smallpox a problem in Europe too? *In Europe during the 18th century, smallpox killed about 400,000 people a year.*

How and when did smallpox come to the Americas? *In 1495, slaves imported by Columbus brought the smallpox virus to the West Indies.*

What percentage of Native Americans are thought to have been killed by the virus? *The virus killed 60 to 80% of the Native Americans.*

How did this virus help bring down the Aztec Empire? *In 1520, Hernan Cortes fought and lost to the Aztec Indians. One of Cortes' men had smallpox. Cortes and his army left in defeat; the virus did not. When Cortes came back one year later, many of the Aztecs were already dead from the disease and he was able to defeat the sick, dying, and weakened people who were left.*

Smallpox is considered to be an eradicated virus. What does that mean? *A disease is eradicated when there are zero cases being transmitted worldwide and treatment measures are no longer required.*

How was smallpox eradicated? It was eradicated by a worldwide vaccination campaign led by the World Health Organization (WHO).

Activity

Organ Systems Work Together

This section requires teacher guidance. It is intended as a review. Now that students are familiar with the major organ systems of the human body, it is important that they step back and look at how these systems all work together in concert.

Answers

Digestive System 1. This system takes food and turns it into molecules your cells need to grow and to make energy. Since all the organs in your body are made from cells, this system powers the other eleven organ systems. This system works directly with your circulatory system when blood carries food molecules to your cells.

Urinary System 2. This system controls the amount of water in your body. Chemical reactions in your body occur in water. Chemical reactions occur in every organ system in your body. Therefore, this system works with every organ system in your body to make sure they have the water they need. This system also removes waste generated by the other organ systems.

Nervous System 3. This system connects your brain to the other organs in your body. It controls the functioning of all the other organ systems.

Endocrine System 4. This system is in charge of maintaining homeostasis. It works closely with the circulatory system, making and sending hormones to the cells in your body. Hormones tell the cells in the organs of each organ system what to do to keep your body within the narrow ranges of homeostasis. Every system in your body participates in maintaining homeostasis; therefore, every system works with this one.

Lymphatic System 5. This system connects the organs of your immune system. It works with your immune system to keep your other organ systems from getting sick.

Circulatory System 6. This system connects all the organ systems to one another. It exchanges oxygen and carbon dioxide with your respiratory system for the other systems. It transports food and water molecules for your digestive and urinary systems to the other systems. It carries hormones for your endocrine system to the other systems.

Skeletal System 7. This system gives you and your organ systems support and protection. This system is necessary for motion, which allows you to get things you need like food and water. This system is where blood cells are made. Red blood cells are the part of your blood that carries oxygen and carbon dioxide back and forth to and from your other systems. Your white blood cells are the pathogen fighters in your body.

Integumentary System 8. This system protects all the other systems by enclosing them in a protective case. It helps maintain body temperature and water homeostasis, which is important to all your organ systems.

Reproductive System 9. Without this organ system you would not be. It is because of this organ system that you have the other eleven organ systems.

Respiratory System 10. Every cell in every organ in every organ system relies on this organ system. No cell, tissue, organ, or organ system can function without oxygen or with too much carbon dioxide. This system takes oxygen into your body and lets carbon dioxide out of it. Your circulatory system takes care of the rest.

Immune System 11. Every organ system in your body is counting on this organ system to keep them healthy by protecting them from pathogens.

Muscular System 12. This organ system is all about movement. Your brain says go, and this organ system does. It moves the other organ systems where they need to go to get what they need to function.

Lab 2

Putting Them All Together

This section requires teacher, parent, or fellow student involvement. It is a continuation of the theme from the text.

Show What You Know

Immune and Lymphatic Systems

Multiple Choice

1. Mucus and stomach acid prevent pathogens from getting into the rest of your body. What else is a barrier for infection? *Skin*

2. The first time your body encounters a pathogen, *the pathogen has time to multiply, until your immune system figures out how to kill it.*

3. Vaccines work by *exposing your immune system to a less toxic form of the pathogen.*

4. To be immune means *your body recognizes a pathogen and because of that it can no longer make you sick.*

5. The main function of memory lymphocytes is *remembering pathogens your body has encountered before.*

6. White blood cells kill pathogens by *engulfing them or breaking them apart.*

7. When pathogens make you sick, you can get a fever. What causes fevers? *Your immune system is fighting the infection by cooking the pathogens.*

8. The organs of the immune system are connected by *lymphatic vessels.*

9. This organ has white blood cells that filter pathogens as they enter through your mouth and nose. They can become infected and are sometimes removed. *Tonsils*

10. Where are memory lymphocytes made? *Peyer's patches*

Story

You wake up and think about a dream you had last night. You wonder what part of your <u>brain</u> makes your dreams. Your <u>nervous</u> system receives a signal telling you to get up because your <u>urinary</u> system just sent it the message that you have to pee. Your <u>bladder</u> must have filled up with urine while you were sleeping. While you are in the bathroom your <u>stomach</u> starts to growl; your <u>digestive</u> system must be sending it signals, saying "Feed ME!" Your <u>endocrine</u> system is using the <u>homeostatic feedback mechanism</u> to make you hungry because your cells need <u>glucose</u> for cellular respiration so you can have enough energy. You make a healthy breakfast because you know your <u>cells</u> are going to need lots of molecules to build more, because you are growing. In fact, you are probably making more <u>blood</u> right now in the <u>spongy bone</u> part of your bones, so your <u>circulatory</u> system can carry food molecules throughout your body to those cells. You put your hand in the middle of your chest and feel your <u>heart</u> pumping that blood where it needs to go.

It is raining. After breakfast you go outside and feel rain as it hits your <u>skin</u>. Your integumentary system must have a lot of <u>nerves</u> running through it for you to feel rain hit it. After catching some raindrops on your tongue, you breathe air deep into your <u>lungs</u> so your <u>respiratory</u> system can have lots of <u>oxygen</u> for all the puddle jumping you have planned. Your <u>muscular</u> system gets ready as you tense your <u>muscles</u> so you can jump from puddle to puddle. It's a good thing you have your <u>skeletal</u> system to support it all or you would be a puddle yourself. Oops, you slip and cut yourself. You can almost feel your <u>immune and lymphatic</u> systems getting ready to do battle with <u>white blood cells</u> rushing to the site.

Lesson Review

Immune and Lymphatic Systems

Your integumentary, respiratory, and digestive systems keep most pathogens out. When pathogens get inside you, your immune system and lymphatic system contain them and kill them.

Immune and Lymphatic Systems

Purpose: to keep your body from getting sick

The primary organs are your lymph nodes, lymphatic vessels, tonsils, thymus, spleen, and Peyer's patches.

How your immune and lymphatic systems find, destroy, and recognize pathogens:
1. A pathogen your body has never encountered before gets inside your body.
2. The pathogen begins to reproduce exponentially.
3. White blood cells carried by your lymph and blood attack and kill the pathogens.
4. A fever kills pathogens by "cooking" them to death.
5. The next time your immune system encounters this same pathogen, white blood cells called memory lymphocytes recognize and destroy the pathogen before it can make you sick.

Unit IV: Anatomy and Physiology

Answer Key Unit Exam Chapters 11–19

The exam for Unit IV is found in the appendix of the student Workbook.

1. **Multiple Choice** (2 points each = 40 points)

The organ(s) in plants where photosynthesis takes place: *Leaves*

The process plants use to move food from the site where it is made is called *translocation.*

Your ears, eyes, mouth, and nose work most closely with which organ system? *Nervous*

The birth process has three stages. The order they follow is *labor, delivery of baby, delivery of placenta*

Your skin and white blood cells fight infections together. How? *Your skin prevents most germs from getting in, but when they do get in, the white blood cells destroy them.*

When a plant breaks out of the seed and begins to grow, it is called *germination.*

When a mosquito bites your hand, nerve cells in your hand send a signal telling your brain to smack it using *electrical signals.*

Where are blood cells made? *Inside your bones*

When you bend your knee, *tendons* connecting your bones and muscles push and pull them, while the *ligaments* hold the bones together, and the *cartilage* helps the bones glide against each other.

Sperm are produced in male organs called *testes.*

During ovulation, *an egg is released into the fallopian tube.*

The chemical messengers that send information back and forth between your organs are *hormones.*

The part of your eye where images are formed is the *retina.*

Blood is what type of tissue? *Connective tissue*

Xylem and phloem are what type of tissue? *Vascular tissue*

Skeletal muscles are *voluntary muscles.*

Bile is a chemical that *breaks down fat.*

Insulin decreases the amount of glucose in your blood, and *glucagon* increases the amount of glucose in your blood.

Once white blood cells have encountered a pathogen, the pathogen can't make you sick ever again. This is called being *immune* to that pathogen.

When you cut yourself, particles in your blood called platelets produce a chemical, which forms a net that traps blood cells and plasma, and forms a clot. This chemical is called *fibrin.*

2. Flower parts. Label the flower (2 points each = 16 points). Use the flower to diagram the three steps of the fertilization process for plants. (2 points each = 6 points).

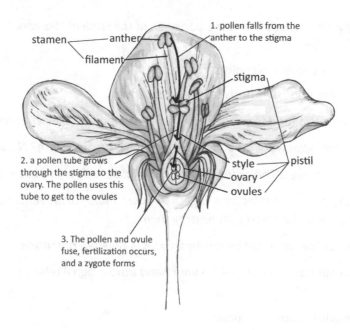

3. Circulation. Blood flows through arteries, veins, and capillaries. What is the difference between them? Fill in the blanks below. (2 points each = 6 points)

Arteries carry blood away from your heart to the rest of your body.

Capillaries are one cell thick, which allows the transport of small molecules across the walls of the membrane.

Veins carry blood to your heart from the rest of your body.

4. Homeostasis. What is homeostasis? (2 points) *Homeostasis is the maintenance of an organism's internal conditions within a normal range.*

Describe how plants use the process of transpiration to maintain homeostasis. (2 points) *Transpiration is the movement of water from the roots to the leaves in plants. It is used to maintain the water balance in plants.*

Plants need to control their water balance because (1 point each = 3 points)

1.) Water is needed for photosynthesis.

2.) Water fills plant vacuoles so they are turgid; turgid cells help plants support themselves.

3.) Chemical reactions take place in water.

Your kidneys are important for your body maintaining homeostasis. What are your kidneys responsible for? How does that relate to transpiration in plants?

Kidneys are responsible for maintaining water balance. (2 points)

Transpiration in plants and kidneys in people both maintain water balance. (1 point)

5. Matching. Write the letter in the space matching each organ system with the organs in it. (1 point each, 11 points total)

A.	Nervous system	_B_	skin, nails, hair
B.	Integumentary system	_K_	lymph nodes, lymphatic vessels, tonsils, thymus, spleen, Peyer's patches
C.	Digestive system	_I_	muscles
D.	Urinary system	_A_	brain, spinal cord, nerves
E.	Reproductive system	_G_	heart, arteries, veins, capillaries
F.	Endocrine system	_C_	salivary gland, esophagus, stomach, small intestine, liver, pancreas, gallbladder, large intestine, rectum, anus
G.	Circulatory system	_J_	bones, ligaments, cartilage, tendons
H.	Respiratory system	_D_	kidneys, ureters, bladder, urethra
I.	Muscular system	_F_	hypothalamus, endocrine glands, thymus, pancreas, testes OR ovaries
J.	Skeletal system	_E_	testes, scrotum, urethra, penis OR ovaries, fallopian tubes, uterus, cervix, vagina
K.	Immune and lymphatic systems	_H_	larynx, trachea, bronchi, lungs

6. Organ systems. Write the name of the organ system from the list that best matches it with the stated purpose. (1 point each = 11 points total)

Muscular system - Works with the skeletal system so you can move

Reproductive system - Makes us

Skeletal system - Gives support and shape; works with the muscular system so you can move

Integumentary system - Protects the inside of the body, to send sensory information about touch to the brain

Immune and lymphatic systems - Keep your body from getting sick

Nervous system - Controls your other systems; stores memories and allows for thought

Urinary system - Removes waste; balances the amount of fluid in your body

Circulatory system - Carries blood throughout your body

Digestive system - Takes food and mashes it down into molecules that cells can transport across cell membranes

Endocrine system - Produces hormones

Respiratory system - Delivers oxygen from the air to your blood

7. Extra Credit. With a partner, use your body to show the approximate location of each of your organs. One point for each organ you correctly locate and name. *Use your student's colored diagrams in his text for answers.*

(/100) x 100 =   +   extra credit points

Unit V: Evolution
Chapter 20: A Story of Luck

WEEKLY SCHEDULE

Two Days

Day 1
❑ Lesson
❑ Lab

Day 2
❑ MSLab
❑ FSS
❑ Lesson Review
❑ SWYK

Three Days

Day 1
❑ Lesson
❑ Lab

Day 2
❑ MSLab
❑ FSS

Day 3
❑ Lesson Review
❑ SWYK

Five Days

Day 1
❑ Lesson

Day 2
❑ Lab

Day 3
❑ MSLab

Day 4
❑ FSS

Day 5
❑ Lesson Review
❑ SWYK

FSS: Famous Science Series
MSLab: Microscope Lab
SWYK: Show What You Know

Introduction to Unit V

One of the fundamental questions to all of biology: How did there come to be so many different types of organisms? The process that has led to the great variety of life is evolution. Evolution is covered in four chapters:

Chapter 20 - Gives a timeline of important evolutionary events.

Chapter 21 - Explains how evolution occurs. It details the processes that lead to speciation.

Chapter 22 - Explains the evidence used to show that evolution is the process responsible for the diversity of life.

Chapter 23 - Explains the processes used to determine the age of fossils.

Lab Preparation: Collect dead winged insects from various species for the microscope lab in this chapter.

Introduction

Chapter 20 explains evolution from the perspective that it is luck, and not superiority, that decides who survives and who doesn't. You might wonder why the phrase "survival of the fittest" is not used in the discussion on natural selection in this evolution unit. Scientists have gotten away from this misleading phrase and have come to understand that luck, or accident, plays the major role in which traits are the most beneficial and give a species the best chance of surviving during extinction events. Evolution occurs through the act of genetic change. Genetic change occurs through mutations and crossing over; these two things are random. So any mutations or crossing over that benefits an organism are random events. The organisms didn't do anything to make these events happen, and they are lucky that they occurred. Sometimes these changes are neutral or even harmful until conditions change, then they become beneficial, which is lucky.

Let's look at the example of bacteria in a petri dish. Let's say there are 100 bacteria in the dish. At the beginning of the experiment, parent cells were genetically identical. These bacteria are being tested with various antibiotics to see which are most effective. This is stressful to the bacteria. These types of stressors will lead to mutations. One of the bacteria was "born" with a mutation in its DNA making it resistant to the antibiotics being tested. The next time these bacteria are exposed to antibiotics, only one will survive, the one that is resistant to the antibiotics. It will reproduce exponentially, and very quickly all the bacteria in the petri dish will be resistant to the antibiotics. This is evolution and is an example of the fittest individual surviving. But that fitness was a result of luck. It was just luck that that bacteria evolved to become resistant to those antibiotics. This logic applies to all cases presented in this chapter.

Learning Goals

- Become familiar with the timeline of major evolution events.

- Learn the predominant theories explaining how the major events in evolution happened.

- Understand that the course of evolution is random, and that being in the right place at the right time, luck, is the main thing affecting who survives and who doesn't, and which traits become predominant.

- Recognize that biological evolution occurs to a population NOT an individual.

- Recognize that there has been evolution from simpler organisms to more complicated organisms.

- Understand that the earth is constantly changing.

- Learn what the word "extinct" means, and recognize it is a normal part of a changing world.

- Learn the definition of the term "geologic time."

- Learn that scientists have divided geologic time into specific time spans.

Extracurricular Resources

Books

Born With a Bang: The Universe Tells Our Cosmic Story: Book 1 (Sharing Nature With Children Book), Morgan, Jennifer

From Lava to Life: The Universe Tells Our Earth's Story (Sharing Nature With Children Book), Morgan, Jennifer

Evolution of Life, Understanding Science and Nature, Time-Life

Online

Visit Pandia Weblinks for videos and websites recommended for this chapter:

www.pandiapress.com/weblinks-biology2

Lesson

Right Place, Right Time

The lesson is written in a timeline fashion that is meant to fit neatly with the lab. The theories presented for different evolutionary events are not the only theories for how these events happened. They are the most predominant theories at this time. The fossil record clearly demonstrates the continuation of life, filled with changing forms, new organisms, and the evolution toward increasing complexity.

Lab

Evolution: A Timeline

When you look at all the numbers in this lab it might scare you away. It is all measuring, though. Take your time with it. After making the first set of measurements, the eons, you will be a pro. Use the timeline to refer to as you go through this unit. The purpose of this lab is to visually display the order of evolutionary events and demonstrate how long evolution takes. Sometimes people will say that they do not believe in evolution because they do not see it happen. That is because for some types of organisms, evolution occurs over long time periods. It can be a slow process for a population to accumulate enough viable mutations that they evolve into a different species. When people say evolution does not occur in the short term, however, they are completely ignoring bacteria.

The amount of time this lab takes can vary depending on the markers you use. I have provided paper markers that your student cuts out and glues to cardboard or poker chips. But if you want to turn this into a more permanent timeline, disregard the paper markers and make your markers by hand, painting images and words on objects like flat rocks, strips of wood, or those made out of clay. You could also use other objects for the markers such as toy animals, dinosaurs, insects, plants, etc. If you make your markers by hand, this lab can take the better part of a day or even a week. When I did this lab with my son, we made our timeline in the backyard and painted the markers on rocks. The timeline is still there as I write this, a year later. This is a good lab to do with your students. Review the events and what led to the events as you place the markers in their spot.

On the next page is an example of a scaled timeline based on inches, not feet. This is provided to assist you in placement of the markers. The example has been split to fit it on the page; your timeline should be one continuous line. If your timeline is measured using feet, as recommended, the evolution markers will not be bunched up and overlapping as they are on the example.

Math This Week

Calculating and measuring the distance for the placement of the markers.

4.5 billion years ago

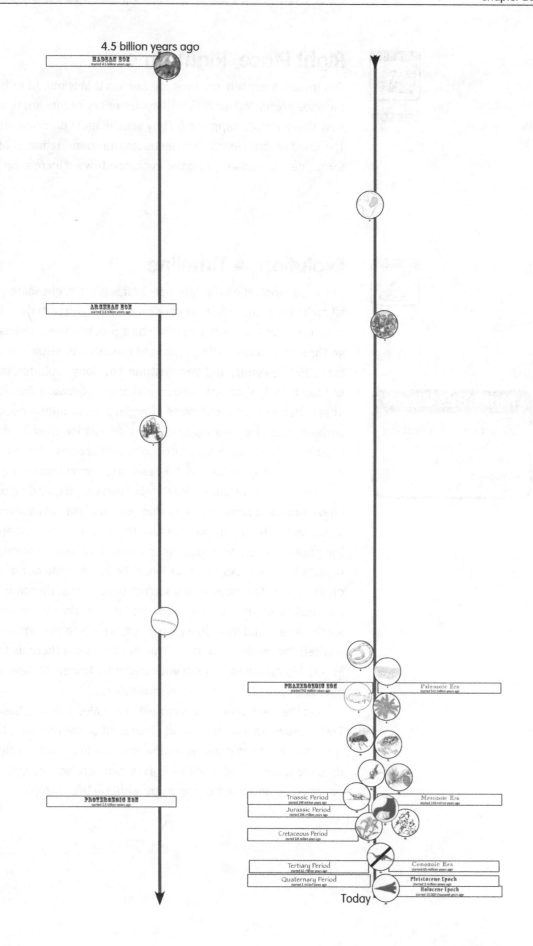

HADEAN EON
started 4.5 billion years ago

ARCHEAN EON
started 3.8 billion years ago

PHANEROZOIC EON
started 542 million years ago

Paleozoic Era
started 542 million years ago

PROTEROZOIC EON
started 2.5 billion years ago

Triassic Period
started 248 million years ago

Mesozoic Era
started 248 million years ago

Jurassic Period
started 206 million years ago

Cretaceous Period
started 144 million years ago

Tertiary Period
started 65 million years ago

Cenozoic Era
started 65 million years ago

Quaternary Period
started 2 million years ago

Pleistocene Epoch
started 2 million years ago

Holocene Epoch
started 10,000 thousand years ago

Today

Pandia PRESS

Microscope Lab

On the Wings of a Bug

This is a fascinating lab. I was amazed when I began looking at insect wings with my microscope to see the variety. To collect an assortment of insects, I looked in the windowsills of my neighbor's house and my house. I found a gnat, a housefly, a wasp, and a dragonfly. My observations:

Gnat: 1 pair of wings, had a lot of dots on the wings
Veins: had small veins running through its wings
Hair: long hairs on the wings; this was the smallest insect, but it had the longest hairs

Housefly: 1 pair of wings
Veins: were darker than the veins of the other 3 insects
Hair: more hair than the gnat, but less hair than the wasp. The hairs on the housefly's wings were short and bristly looking.

Wasp: 2 pairs of wings
Veins: it looked like the veins on the edges of the wasp's wings were different from those on the other parts of the wings; they were thicker and darker
Hair: the wasp had the most hair of the 4 insects, but there were no hairs on the edges of the wings

Dragonfly: 2 pairs of wings
Veins: had big veins, the biggest and most of the 4 insects
Hair: no hair; this was the only insect with no hair

Famous Science Series

The Burgess Shale

The running parallel through this chapter is luck. The discovery of the Burgess Shale fossils was very lucky. Since the time of the discovery, fossils from the same time period and of the same and similar organisms have been found at sites all over the globe.

What did Charles Walcott see? *The rock split in two and inside was a fossil.*

Where is the Burgess Shale site? *Today the Burgess Shale is 8,000 feet above sea level in the Canadian Rockies. 505 million years ago, this site was below sea level.*

How many types of fossilized organisms have been found at this site? *170*

List at least two things that make the Burgess fossils special.

1. *Most fossils are from the hard parts of organisms: shells, bones, and teeth. At the Burgess Shale site, the soft tissue of the organisms fossilized as well as the hard parts.*
2. *Some of the fossils from this site are of fantastic-looking organisms that seem like they could be evolutionary dead-ends. Others could be the distant relatives of organisms alive today.*

3. The entire group of fossilized organisms taken together gives insight into what a community filled with multicellular organisms must have been like early in the evolution of such complex systems.

What happened to the organisms over 500 million years ago that created the Burgess Shale site? *An underwater avalanche of fine mud fell on top of he organisms, burying them alive.*

Show What You Know

Evolution: A Story of Luck

Multiple Choice

1. Life on Earth evolved *before 3.5 billion years ago.*
2. Three billion years ago, photosynthesizing organisms evolved. They spewed a waste product into the water and air. What was the waste product? *Oxygen*
3. From question 2: The evidence for this is found all over the world in *banded iron formations.*
4. Biological evolution *happens to a group.*
5. The oldest fossils are *stromatolites.* They are *3.5 billion years old.*
6. An example of a population is *all the Emperor penguins that live and breed together in a small area in the Weddell Sea region of Antarctica.*
7. The Endosymbiotic Theory explains how *eukaryotic cells evolved.*
8. Geologic time *is the time from when the earth was formed to the present day.*
9. The creation of an ozone layer was important because *it created a protective shield from the sun's rays so that organisms could colonize land.*
10. A group of organisms called archosaurs gave rise to: *All of the above*
11. What trait did plants and animals have at the time that made the colonization of land possible? *A protective outer layer*
12. Mass extinctions occur when *a large number of species die within a short period of geologic time.*

Questions

1. Dinosaurs roamed the earth about 1.65 million years. Sixty-five million years ago an asteroid hit Earth at the same time that massive volcanic eruptions were occurring. Explain how these two catastrophes led to the mass extinction of the dinosaurs.

   Both the asteroid and volcanic eruptions threw dust into the atmosphere, which blocked out sunlight. This caused temperatures on Earth to drop and interrupted photosynthesis, which would in turn interrupt the global food web as plants became less productive.

   In addition, the asteroid impact triggered tsunamis and earthquakes. The volcanic eruptions released poisonous gases into the atmosphere and caused massive wildfires. All of these environmental catastrophes led to the extinction of dinosaurs.

2. There were several evolutionary steps going from simple prokaryotic organisms to complex multicellular eukaryotic organisms. Give a step-by-step explanation for how this could have occurred.

   1. *Bacteria that evolved into the organelles mitochondria and chloroplasts came to live inside larger simple cells. Cells that took in photosynthesizing bacteria could make their own food. Cells that took in bacteria that became mitochondria could use oxygen and food to make energy. Cells with this symbiotic relationship evolved into the first eukaryotic cells.*
   2. *These unicellular eukaryotic cells grouped together.*
   3. *Over time, the cells began to work together as individual cells in the group began to specialize. This benefited the whole group as some cells collected food, some cells transported food, and some cells made energy.*
   4. *These groups became the first multicellular eukaryotic organisms.*

3. When dinosaurs went extinct, another group of animals, mammals, possessed a number of traits that made it possible for them to survive the environmental catastrophes occurring on Earth. What were these traits, and how did they benefit mammals?

   1. *Fur – temperatures were low on Earth and fur is warm*
   2. *They were small in size – the global food web was interrupted because less sunlight was reducing plants' ability to photosynthesize. Organisms that were small needed less food.*
   3. *They were omnivores. Omnivores eat many different types of food. When food is scarce, omnivores are more likely to find food to eat.*
   4. *They were burrowing animals – burrowing animals were able to go underground to escape the environmental disasters happening above ground.*

Lesson Review

Evolution: A Story of Luck

The lesson review for this chapter is very short. For the main part of your review, go over the timeline from the lab. If you made a large timeline, walk along it and discuss what led to the evolutionary events on the timeline. Ask lots of questions.

Evolution happens through genetic change.

Biological evolution = genetic change that is inherited within a population

Biological evolution does not occur to one individual; it occurs to a population.

Extinction = all the populations of one species dies

Mass extinction = large numbers of species die within a short period of geological time

Geological time is LONG, it spans from today to 4.54 billion years ago.

Unit V: Evolution

Chapter 21: How

Two Days

Day 1
- ❑ Poetry
- ❑ Lesson
- ❑ Lab

Day 2
- ❑ MSLab
- ❑ FSS
- ❑ Lesson Review
- ❑ SWYK

Three Days

Day 1
- ❑ Poetry
- ❑ Lesson
- ❑ Lab

Day 2
- ❑ MSLab

Day 3
- ❑ FSS
- ❑ Lesson Review
- ❑ SWYK

Five Days

Day 1
- ❑ Poetry
- ❑ Lesson

Day 2
- ❑ Lab

Day 3
- ❑ MSLab

Day 4
- ❑ FSS

Day 5
- ❑ Lesson Review
- ❑ SWYK

Introduction

The purpose of Chapter 21 is to explain how evolution happens.

Learning Goals

- Understand the scientific definition of "theory," and recognize that a scientific theory is a work in progress.

- Learn the steps in the process of evolution.

- Learn the causes of genetic variation.

- Learn how natural selection and genetic drift affect the prevalence of traits in populations.

- Learn how speciation occurs.

Extracurricular Resources

Books

Evolution Revolution, Winston, Robert

Evolution (DK Eyewitness Books), Gamlin, Linda

The Leakeys: Uncovering the Origins of Humankind, Poynter, Margaret

Evolution, Silverstein, Alvin

Mammals Who Morph, Morgan, Jennifer

Online

Visit Pandia Weblinks for videos and websites recommended for this chapter:

www.pandiapress.com/weblinks-biology1

Poetry

A Recipe for Making Something Different

This poem was first-place winner in the 2012 Lyric Division of Spellbinder Books' poetry contest. I wrote it to be used as a teaching poem, and recommend reciting it throughout the week. As this chapter demonstrates, the explanation isn't quite as simple as the poem's refrain indicates, but as a bare-bones answer the refrain is correct. The recipe for speciation is isolation, variation, and lots of time.

Lesson

How Evolution Happens

Over 1.75 million species of organisms have lived on Earth. Evolution is the process that has led to the great diversity of life. That is a fact; it is not a theory. What is a theory is exactly how the process of evolution happens. For example, is it a fact that the Endosymbiotic Theory is definitely how eukaryotic cells evolved? It is not a known fact. It is the best explanation for the data known at this time. That is what makes it a theory not a fact.

In this chapter when explaining overproduction I state that, "...some mice will have a higher survival rate than others will. It is part luck and part genetics." Remember, the combination of genes an organism gets from its parents is random. Which genes end up in which gamete during meiosis, and then end up in the sperm and egg that make the zygote, is a random process. If one sibling has a genetic makeup that increases its chances of survival over another sibling, this is just by random chance, also known as luck.

Genetic recombination, crossover, was first introduced in Chapter 8. It is not a type of mutation. The genes do not change their chemical composition, as happens in mutations. They shuffle and pair into different combinations.

Lab

Natural Selection

This lab asks the question: If I was a mouse, what color would I want to be? The answer is whatever color the ground is. This is a very good lab for demonstrating natural selection. Be careful that your students do not look at the pompoms while you are hiding them. Keep track on the lab sheet as you go along.

This lab is a nice one for a formal lab report. I have provided a sample.

Possible Answers

Your answers will vary from mine for these questions. The following answers are from a student who completed this experiment outside on sandy ground.

Hypothesis: *I think I will select the most black mice and the least gray mice.*

Round 1 – color of first mouse caught: *Brown*

Round 2 – color of first mouse caught: *White*

Round 3 – color of first mouse caught: *Black*

			black	brown	gray	spotted	white
Round 1		caught	2	2	0	4	3
		left	3	3	5	1	2
		added	3	3	5	1	2
	total mice in next round		6	6	10	2	4
Round 2		caught	1	3	1	1	4
		left	5	3	9	1	0
		added	5	3	9	1	0
	total mice in next round		10	6	18	2	0
Round 3		caught	5	4	6	2	0
		left	5	2	12	0	0

Math This Week

1. Pompoms are counted.
2. Students must keep track of the color of pompoms.

Did you catch the same color mouse first every time? What do you interpret from that? *No, the first mouse I caught was the most random because it was the one closest to me. In the next two rounds the first mouse I caught was the one that stood out the most because of its color.*

Which colored mouse was best adapted for the environment? Which colored mouse was the worst adapted for the environment? *The gray was the best adapted; the white was the worst adapted closely followed by the spotted.*

Based on the results from this experiment, why do you think the mice in the wild are brown or brownish-gray in color? Use the term *natural selection* in your answer. *There is natural selection for hair color. In nature (but not in my experiment for brown), mice that blend in best with their environment are less likely to get eaten. There is selection against fur colors that stand out, like white.*

In terms of evolution, fitness is defined as the ability to produce offspring. Which fur color results in the best fitness for the mice? *gray*

If there is continued selection for and against certain fur colors, what do you think will be the color of the mice in this population? *There will be more gray mice but maybe some black ones too.*

What happened to the mouse that had the best fur color once it became more numerous? *It started to get eaten more because there were fewer other color choices. But it still increased a lot in number. If the experiment were to go another round, there would be 24 gray mice in the next round. That is almost 5 times what I started with.*

Chapter 21: Lab Report

Name: _Oliver Smith_ **Date:** _12/30/2012_

Title: _Natural Selection_

Hypothesis

I think I will select the most black mice and the least gray mice.

Procedure

I had my mom scatter 5 pompoms each of several different colors over a 2 to 3 square meter area at a local playground. I did not look while she was doing this. After she scattered the pompoms, she told me to open my eyes, then try to pick up as many pompoms as possible in 30 seconds. For each pompom I did not pick up, I added another one of that color to be hidden in the next round. I did this a total of three times.

Observations

Round 1 – the first pompom I picked was brown. I picked up the most spotted ones, 4, and no gray ones.

Round 2 – the first pompom I picked was white. I picked up the most white ones, 4, and only 1 gray one, 1 black one, and 1 spotted one. There were no more white pompoms for the next round, I picked them all up.

Round 3 – the first pompom I picked was black. I picked up the most gray ones, 6, and the least spotted ones, 2, but that was all of the spotted ones.

Results and Calculations

See data table from lab sheet for calculations
By the end of round three there were 12 gray pompoms, 5 black pompoms, 2 brown pompoms and no white or spotted pompoms.

The pompom color of the first pompom picked each round was only a good indicator of which color would be the one I caught most for round 2.
Round 1: brown picked 1st; I caught 4 spotted but only 2 brown total
Round 2: white picked 1st; I caught 4 white pompoms, all there were
Round 3: black caught 1st; I caught 6 gray and 5 black, but there were more gray pompoms

Conclusions

In conclusion, the gray did blend in the best as I thought in my hypothesis, but black blended in better than I thought it would. White and spotted pompoms showed up the best on the surface where they were scattered, and were the easiest to find when I was racing around trying to pick up as many pompoms as possible in 30 seconds. This experiment makes it easy to understand how there could be natural selection for fur color in nature. Animals that stand out are easier to quickly find the way a predator would need to do when catching its prey. That color would be selected against and would become less common and possibly nonexistent in nature.

Microscope Lab

Function and Form

Why are there so many different colors and forms of hair/fur? That is just the type of question Darwin and his peers were asking when they began to look for a scientific explanation for the many variations. Try to look at as many different types of fur as you can. You could add a slide of a feather to this lab too. Feathers in many ways function the same for birds as fur does for mammals.

Possible Answers

Your answers will vary from mine for these questions. The following answers are from a student who completed this experiment.

What two fur samples from different types of animals do you think will look the most alike with a microscope? Why? _Cat hair from a fuzzy cat and sheep hair because they have the most similar texture._

What two fur samples from different types of animals do you think will look the most different with a microscope? Why? *Dog hair and guinea pig hair because they look and feel the most different from each other.*

Type of Fur	Appearance Description	Comments: Compare/Contrast
Dog, border collie	coarse, thick black hair	
Dog, white mix breed	white hair on outside, black inside	texture was like people's hair
Cat	thin and fuzzy	
Sheep	super thin and fuzzy	looked almost identical to the cat hair
Guinea pig #1	straight, thick, white, and black	I could see a large cuticle
Guinea pig #2	straight red	thinner texture than guinea pig #1
Person #1	straight, brown, thin	
Person #2	curly, blond, thick	This was the thickest of all the fur samples

What two fur types looked the most alike with a microscope? Why? *Cat and sheep; they were hard to tell apart they were so similar. They were the same thickness in diameter and both had similar-looking cuticles (the part around the outside of the hair).*

What two fur types looked the most different with a microscope? Why? *Guinea pig and sheep; the guinea pig hair was thick and coarse and the sheep hair was thin and fuzzy; the guinea pig hair had a thicker cuticle around it.*

Were you surprised by anything you saw? If so, what? *My white-haired dog; he looks totally white, but under the microscope his hair is black on the inside and white on the outside.*

Famous Science Series

Evolution Act 1: First Theories

My goal with this Famous Science Series is to set the stage to get students thinking about people's mindset and where the science was at just before the time Darwin proposed his theory of evolution.

In 1809, Jean-Baptiste Lamarck proposed a theory of evolution. What was his theory and was it correct? *Lamarck believed the traits an organism acquired during its lifetime were passed to the organism's offspring. He proposed that if a species used an organ more than they had in the past, that the size of the organ would*

increase, such as the neck of a giraffe. The longer the giraffe's neck was, the easier it would be to feed from leaves at the tops of trees. These longer-necked giraffes would give birth to even longer-necked giraffes. His theory was incorrect and had its problems. Chief among them was the lack of supporting data. Nevertheless, it was a start and it got people thinking.

In 1788, James Hutton, a Scottish farmer and geologist, put forth a theory called uniformitarianism. What was the theory and was it correct? *He proposed that:*
- *The earth was much older than the 10,000 years people at that time believed.*
- *The earth is constantly changing.*
- *The processes that cause change are slow, but many of them are observable:*
    - *o erosion caused by wind and water*
    - *o volcanic eruptions, bringing melted rock from the center of the earth, which then cools and builds up the earth's surface*
    - *o sediments deposited layer upon layer, becoming rock*
- *These processes have been constant over the life of the earth.*

James Hutton's theory of uniformitarianism is correct.

Show What You Know

Evolution: How

Multiple Choice
1. One bacterium splits into two, then two to four... Soon there are millions. The bacteria run out of food and begin to starve to death. This is an example of: *Overproduction*
2. The case of the peppered moths is a good example of how *natural selection* works.
3. What two mechanisms lead to genetic variation? *Genetic recombination and mutation*
4. Aquatic birds, like ducks, have webbed feet that help them paddle through water. This is an example of *an adaptation.*
5. Dogs and cats are not the same species because *they cannot breed with each other and have offspring.*
6. Which of the following statements are true? *All of the above*
7. This change to an all-green population on the new island is an example of *genetic drift.*
8. Arctic hares have fur that is brown in the summer and white in the winter. If the earth became warmer and all the snow melted in the Arctic, this would be an example of a *beneficial* trait that became a *harmful* trait.
9. Reproductive isolation is necessary for speciation because *gene flow must be stopped between populations for one to evolve into a new species.*
10. Mutations are *random*; selection for the traits they cause is not.

11. What is the name of the process that explains how all the species of organisms have come to be? *Evolution*
12. Overproduction should lead to there being many more organisms alive than the earth can support. What are the controls on overproduction? *All of the above*

Essay Answers

1. This essay takes some thought but it can be done. Here is an example:
Many years ago a terrible hurricane struck the Island of Mythical Creatures and a small group of sealocrabs became separated from the main population, when they were swept out to sea and marooned on a small island. The main island was too far for them to swim back to it. There was <u>genetic variation</u> in this small group, which led to the <u>natural selection</u> for certain <u>adaptations</u> that were favored in their new environment. The beneficial adaptations were selected for and increased in the group. Because the group was small, there was <u>genetic drift</u> for some <u>traits</u>. The group was <u>reproductively isolated</u> from its parent population for a long <u>time</u>. Because of <u>genetic recombination</u> and <u>mutation</u>, new traits <u>evolved</u> in the group. Eventually, there was <u>speciation</u> with the <u>evolution</u> of the sealocrab into the aquanotic.

2. *The cell theory was established in 1855.*
1665, Robert Hooke discovers squarish-looking structures in a specimen of cork with his microscope. He names these structures *cells*.

1674, Anton Van Leeuwenhoek is the first person to see a live cell with his microscope.

1831, Robert Brown discovers the cell nucleus.

1839, Thoeder Schwann and Matthias Schleiden propose a theory stating that all living things are made of one or more cells. Schwann and Schleiden perform many different experiments before proposing their theory. They conduct many experiments after their theory is proposed. *They called it a theory, but they are the only two researchers who have tested it, so it is not a theory yet.*

1855, Rudolf Virchow proposes that every cell comes from another cell. This is added as a part of the cell theory. He performed many different experiments before proposing this. He conducts many experiments after his theory is proposed. *This is the point where the cell theory becomes a legitimate scientific theory.*

1855 to present day, the cell theory has been tested many times by many different researchers. Their results have confirmed the cell theory.
The work Virchow is doing shows that different researchers are testing the findings of Schwann and Schleiden. Virchow adds to the theory, which is common in science. The continued research on working scientific theories leads to addendum and changes as scientists learn more through their research. The last entry is purposely worded to include the definition of a scientific theory in it.

Lesson Review

Evolution: How

A good review is to walk your timeline and discuss what things might have led to the evolution of types of organisms. For example, ask a question like: If birds evolved from dinosaurs, what kinds of mutations and selection would have led to the first birds?

To be of the same species, organisms must be able to . . .
1. Reproduce
2. Have healthy offspring that can also reproduce. For example, horses and donkeys are not the same species because: They can reproduce BUT their offspring, mules, cannot reproduce; they are sterile.

The Process of Evolution

Overproduction = more are born than can survive

Genetic variation = mutations and genetic recombination leads to different, new, and new combinations of traits

Natural selection = there is selection for and against traits

Genetic drift = in small populations, alleles can become fixed in a population where the allele is the only copy of that allele present in the population

Reproductive isolation = organisms are isolated from each other so they cannot reproduce with each other; gene flow is stopped between groups—this is a necessary condition for speciation

There must be genetic variation to have different traits.

Different traits allow for natural selection and genetic drift.

Natural selection and genetic drift change the traits in populations.

Organisms with different traits can evolve away from each other if there is reproductive isolation.

When enough new traits accumulate, speciation can happen.

Unit V: Evolution
Chapter 22: Evidence

WEEKLY SCHEDULE

Two Days

Day 1
- ❑ Lesson
- ❑ Lab
- ❑ MSLab

Day 2
- ❑ FSS
- ❑ Lesson Review
- ❑ SWYK

Three Days

Day 1
- ❑ Lesson
- ❑ Lab

Day 2
- ❑ MSLab
- ❑ FSS

Day 3
- ❑ Lesson Review
- ❑ SWYK

Five Days

Day 1
- ❑ Lesson

Day 2
- ❑ Lab

Day 3
- ❑ MSLab

Day 4
- ❑ FSS

Day 5
- ❑ Lesson Review
- ❑ SWYK

Introduction

This chapter details the evidence upon which our understanding of evolution is based.

Learning Goals

- Understand the different types of evidence for evolution.

- Learn about fossils, how they form, and the fossil record.

- Learn about the homologies in chemistry, genetics, developmental biology, and anatomy between organisms.

- Learn about the genetic evidence showing evolution.

- Learn about biogeography.

Extracurricular Resources

Books

Charles Darwin, Giant of Science, Krull, Kathleen

Charles Darwin and Evolution, Parker, Steve

Mr. Darwin's Voyage, Altman, Linda Jacobs

Darwin and Evolution for Kids: His Life and Ideas, Lawson, Kristan

Lucy Long Ago, Uncovering the Mystery of Where We Came from, Thimmesh, Catherine

Dinomummy, the Life, Death, and Discovery of Dakota, A Dinosaur from Hell Creek, Manning, Phillip

Fossil Fish Found Alive: Discovering the Coelacanth, Walker, Salley M.

Is There a Dinosaur in Your Backyard?: The World's Most Fascinating Fossils, Rocks, and Minerals, Christian, Spencer

Fantastic Fossils, Shone, Paul

Dinosaurs Down Under and Other Fossils from Australia, Arnold, Caroline

The Tree of Life: Charles Darwin, Sis, Peter

A very interesting work of fiction is *The Kin* by Peter Dickenson. It is lengthy and makes a great read-aloud. The setting is Africa 200,000 years ago. It follows a group of hominids, who already have speech, on a search for a new place to live. They meet other hominid groups, some who have no speech.

Game

"American Megafauna, The Continuing Contest Between Dinosaurs and Mammals" This board game is complicated but fun to play once you figure it out. It is a good game about evolution.

Online

Visit Pandia Weblinks for videos and websites recommended for this chapter:

www.pandiapress.com/weblinks-biology2

Lesson

Evidence of Evolution

Some of the information in Chapter 22 is technical. I recommend going through the chapter with students to ensure that they understand it. Genotype is determined by an individual's chemistry. Phenotype is determined by an organism's genotype and therefore chemistry. I separate the different types of homology from each other, and homology from the genetic evidence for evolution. But if you think about it, and some of your students might, all the evidence is really chemical and genetic. The technological advances in the fields of biochemistry and genetics have drastically increased the amount of chemical and genetic evidence of evolution. Research in the fields of biochemistry and genetics adds information about and evidence for evolution every day.

Lab

Forming Fossils

There is a lot of new material in this lab that I did not discuss in the text. Students will be asked questions on this material in Show What You Know questions and in the exam.

Microscope Lab

Layers

Students need to be careful looking at rocks with their microscope. Try to get a very thin slice of sedimentary rock. If you cannot get a very thin slice, a magnifying glass may give better results.

Famous Science Series

Evolution Act 2: Charles Darwin

What was the name of the ship Darwin sailed on? *The Beagle*

How long was he at sea? *5 years*

What was his job on the ship? *Darwin was the ship's naturalist. His job was to collect specimens, make observations, and keep careful records of what he observed.*

What did he conclude was the answer? *He concluded that natural selection was responsible for the differences in the size and shape of the finches' beaks. He compared the different types of food on the different islands. He realized that there was different selection for the size and shape of beaks, based on the available food. Adaptations that helped finches on one island did not help finches on another island. Over time, the finches on each island had accumulated enough genetic differences that they had evolved into a new species of finch.*

What was the complete title of his book? (It's 21 words long!) <u>On the Origin of Species by Means of Natural Selection, or the Preservation of Favored Races in the Struggle for Life</u>

What is the shortened title most people use? <u>The Origin of Species</u>

How long did it take Darwin to write it? *He thought of the theory in 1838. He started writing his book in 1844. He published it in 1858.*

Why did he finally publish it? *He finally finished the book only when another scientist, Alfred Russel Wallace, wrote a letter to Darwin proposing the same theory.*

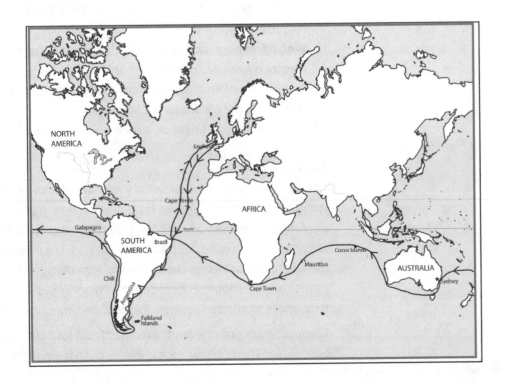

Show What You Know

Evolution: Evidence

Multiple choice

1. A sequence of fossils have been found that show the evolution of horses from fox-sized, forest-dwelling animals with four toes on its front feet and three on its back to the large, one-toed animals we have today. This is an example of *transitional fossils.*

2. In what kind of rocks do fossils form? *Sedimentary*

3. The more closely related two species are, the more homologous their biology and chemistry is. In this sentence the word *homologous* means *similar.*

4. Body parts that share a common ancestor but serve a different function are called *homologous structures.*

5. What body parts are the most likely to become fossils? *The hard parts, like shells, wood, and bones.*

6. What molecule have scientists recovered from some insects trapped in amber and frozen woolly mammoths? *DNA*

7. This is an example of *natural selection.*

8. Humans have intentionally bred milk cows for increased milk production. What is this intentional breeding called? *Artificial selection*

9. Species have traces of DNA from distant ancestors. Sometimes this DNA codes for structures a species does not use, structures that are homologous to functional structures in other species. These are called *vestigial structures.*

10. For many ground-dwelling organisms, the Grand Canyon is a deep, wide, impassable expanse. Rodent species are different from one side of the Grand Canyon to the other, while bird species are the same on both sides. This distribution of animal species at the Grand Canyon is explained by *biogeography.*

11. With permineralization, *an organism's soft tissue is replaced with minerals and creates a fossilized organism that is a rock.*

12. An example of an organism that fossilizes without alteration is *a mammoth that is frozen in the tundra for 24,500 years.*

13. A carbon imprint found on a rock is an example of *All of the above*

Questions

Why do fossils form in sedimentary rock and not the other types of rock? *The intense heat destroys igneous rock, and the heat and pressure metamorphic rock are subjected to destroys any fossils in them. But sedimentary rock usually forms in water where sediments and small bits of rock are pressed down from the weight of the water. This squashes the sediments together, forming rocks. Plants and animals sometimes are caught in between the layers of sediments as the rock forms. Over time, these plants and animals fossilize, becoming rocks themselves.*

The main evidence of evolution comes from four sources. What are they? Give a brief explanation of how each is evidence that evolution occurred.

1. The fossils record shows that
   - Organisms have changed over time
   - They started out simple → increased complexity
   - There are transitional fossils that show evolutionary links

2. Homology – similarities show relatedness. The more similarities there are, the more closely organisms are related. They also show that organisms have evolved from common ancestors.

3. Genetic evidence – Evolution is caused by changes in gene frequency. The selection of traits in nature and artificially by humans leads to changes in gene frequency and evolution. The changes that occur to traits when they are selected for and against are evidence of evolution.

4. Biogeography – explains the distribution pattern of organisms: why they are where they are and aren't in all the places that they could be.

Matching

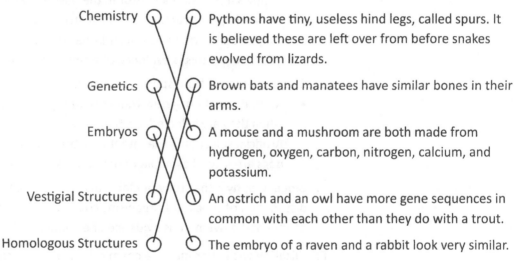

Chemistry — Pythons have tiny, useless hind legs, called spurs. It is believed these are left over from before snakes evolved from lizards.

Genetics — Brown bats and manatees have similar bones in their arms.

Embryos — A mouse and a mushroom are both made from hydrogen, oxygen, carbon, nitrogen, calcium, and potassium.

Vestigial Structures — An ostrich and an owl have more gene sequences in common with each other than they do with a trout.

Homologous Structures — The embryo of a raven and a rabbit look very similar.

Lesson Review

Evolution: Evidence

The main evidence of evolution comes from four sources

1. The **fossil record** shows
   - Organisms have lived on Earth for at least 3.5 billion years
   - Organisms have evolved from simpler to more complex forms
   - Some organisms go extinct and new types of organisms take their place
   - Transitional fossils = fossil with traits that are between one form and another

2. **Homologies** = similarities
   - Homologies in chemistry
     - o organisms are made of cells with the same basic components
     - o organisms use the same energy pathway
     - o organisms are made from the same six elements
     - o all organisms start as one cell
   - Genetic
     - o all organisms have DNA
     - o the closer-related the species are, the more similar their DNA is
   - Embryos
     - o the embryos of very different animal species resemble each other
   - Anatomy
     - o vestigial structures = structures that do not function on a species but are similar to a functioning structure in another species of organism – this shows a common ancestor in the past between these two organisms
     - o homologous structures = similar structure different function, ex. wing, flipper, and arm have similar bone structure but are used for different functions—this is evidence of a common ancestor in the past

3. Genetic evidence from species alive today
   - Natural selection = new traits becoming more common in a population when they are selected for, there is evolution toward that trait
   - Artificial selection = the intentional breeding of organisms for certain traits, dogs are a good example of how this can drastically change an organism

4. **Biogeography** = the study of global distribution patterns of species. The distribution of species of organisms coupled with what we know about continental movement is evidence of evolution

Fossilization with alteration = a chemical process that changes the organism's cells and tissues to a different material
1. **Permineralization** = fossilized parts of the organism mineralize and become rock
2. **Carbonization** = the plant or animal is transformed into a thin film of carbon

Fossilization without alteration = the organism or its parts becomes a fossil without its organic material changing
1. Teeth or bones
2. An organism trapped in amber

Mold fossil = the original remains of the organism are gone, and all that is left is an organism-shaped hole in the rock

Cast fossil = if the mold fills in it is a cast fossil, a three-dimensional representation

Trace fossil = trace fossils are fossils that show evidence, traces, that an organism was there, e.g., nests, footprints, burrows

Unit V: Evolution

Chapter 23: When

WEEKLY SCHEDULE

Two Days

Day 1
- ☐ Lesson
- ☐ Lab
- ☐ MSLab

Day 2
- ☐ FSS
- ☐ Lesson Review
- ☐ SWYK
- ☐ Unit V Exam

Three Days

Day 1
- ☐ Lesson
- ☐ Lab

Day 2
- ☐ MSLab
- ☐ FSS

Day 3
- ☐ Lesson Review
- ☐ SWYK
- ☐ Unit V Exam

Five Days

Day 1
- ☐ Lesson

Day 2
- ☐ Lab

Day 3
- ☐ MSLab

Day 4
- ☐ FSS

Day 5
- ☐ Lesson Review
- ☐ SWYK
- ☐ Unit V Exam

FSS: Famous Science Series
MSLab: Microscope Lab
SWYK: Show What You Know

Introduction

Chapter 23 explains the processes used to date fossils.

Recite the poem "A Recipe for Making Something Different from Me" a couple of times this week.

This is the last chapter in Unit V. There is a Unit V Exam that covers the material found in Chapters 20 through 23, in the appendix of the student Workbook. The answer key is found at the end of this chapter.

Learning Goals

- Learn the Principle of Superposition.

- Learn how rocks are used to date fossils.

- Learn the Principle of Cross-Cutting Relations.

- Learn about index fossils and how they are used for relative dating.

- Learn about relative dating.

- Learn about absolute dating.

- Learn about radiometric dating and how it works.

- Learn about some common radioisotopes used for radiometric dating.

Extracurricular Resources

Online

Visit Pandia Weblinks for videos and websites recommended for this chapter:

www.pandiapress.com/weblinks-biology2

Lesson

Let's Date

Relative dating and absolute dating are both used to determine the age of fossils. Relative dating gives a relative or "about" age for fossils. The methods used for relative dating are the dating of igneous rocks that are close to the sedimentary rocks where the fossils are found, or through the use of index fossils. Radioactive isotopes in fossils 70,000 years old and younger are used for absolute dating with carbon-14. Unstable (radioactive) elements decay into other stable elements at a constant rate. The decay is measured using half-lives, or the amount of time it takes for half of the unstable atoms to decay into stable atoms. The age of older fossils is found using their position in sediment layers and their position relative to index fossils and igneous layers that have been dated using radiometric dating.

Lab

Are We Just Relatively Dating?

This is a logic puzzle. The drawings are based on real fossils found in layers of the Grand Canyon. Some pieces are easier to place than others. If your students are stumped, have them put in the easy strips first.

Possible Answers

1. What information did you use to place the first puzzle piece? *Answers will vary*
2. What is the main type of rock where fossils are found in the Grand Canyon? *Sedimentary*
3. How is this type of rock formed? *Sedimentary rock is formed from sediment that is deposited in layers over time, usually in lakes and oceans.*
4. Locate the Grand Canyon on a map. Why do you think there are both water and land animals fossilized in the Grand Canyon? *The area where the Grand Canyon is has been desert, inland sea, and ocean and lots of things in between. The fossils of water-living organisms were deposited there when it was an ocean or inland sea and land-living organisms were deposited there when the Grand Canyon was not covered in water.*
5. What principle taught in the chapter does this lab demonstrate? *The Principle of Superposition*
6. Scientists are not sure exactly how the Grand Canyon formed. They think it formed through erosion, the movement of tectonic plates, and volcanic eruptions. There was a time millions of years ago when there was no canyon; it was just a flat rocky plateau. At that time there must have been fossils in the layers of the rock. What must have happened to the fossils? How does this show a weakness in using the Principle of Superposition alone to date fossils? *The fossils must have moved. How they moved and where they moved would depend on the process that moved them; for example, washing away with erosion or uplifting through the actions of tectonic plates would change how and where the fossils moved. This could move them to the levels of different layers that were not washed away or to different places altogether and make these fossils useless for dating purposes.*

Microscope Lab

Trees Can Date Too

To get the best view with a microscope, the slice must be smooth. If it is not smooth enough, the magnifying glass will give a better view. Look at the tree rings with a magnifying glass and a microscope. To get a slice of wood, I contacted someone who cuts firewood.

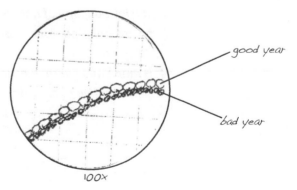

good year

bad year

100x

Comments: *I drew only two rows to represent what I saw. The "good" years had larger cells since the tree received more nutrients in those years. I counted 63 rings total. I estimate the tree was 63 years old. Some rings are very close together and some are far apart. About 25 years before the tree died, there was a group of 9 years that all looked bad except one, based on ring thickness.*

Famous Science Series

Evolution Act 3: After Darwin

Like all scientific theories, the underlying theories proving evolution have been tweaked and changed to fit the results of experiments. When Darwin proposed his theory, he did not know the mechanism for the inheritance of traits. An understanding of this mechanism affected the original theory. Just like the process they describe, there will be continued evolution of the theories behind the process. Evolution happens at the genetic level. With more sophisticated testing methods and with the use of computers, scientists learn more about the genetics of organisms every day. The continued evolution of the theories explaining evolution is not a weakness or a limitation, and it does not disprove evolution. In fact, the continued growth and evolution of scientific theories as scientists better understand a subject is one of the basic tenets and strengths to the field of science. The Modern Synthesis is the best explanation for how evolution works at this time. Who knows where it will be 100 years from now. All we can be sure of is that it will have evolved.

There was one important part missing from the Theory of Evolution as proposed by Darwin. What was it? *Genetics*

When was *The Origin of Species* published? *1859*

When was Mendel's paper on genetics published? *1865*

Look up and give a more detailed explanation of the Modern Synthesis.
The Modern Synthesis explains evolution at the organism, population, species level by incorporating the chemical mechanism that creates the variation, slight differences in the DNA of organisms that creates the variation in the first place.

Many scientists think we are in the midst of a mass extinction right now! What are some of the causes?
1. Habitat destruction, also called loss of habitat. This is mainly through human actions. Two examples are
   - *Humans cut down trees around our houses and in rainforests.*
   - *We spew CO_2 into the atmosphere that heats the planet, which in turn caused polar ice to melt.*
2. The introduction of non-native species of plants and animals; these species have no natural predators. This can wreak havoc on an ecosystem.
3. Pollution
4. Hunting and overharvesting
5. Taking animals for profit

Show What You Know

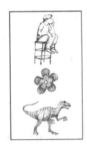

Evolution: When

Multiple Choice

1. The study of the layers of sedimentary rock layers is called *stratigraphy.*
2. There are 1,000 carbon-14 atoms in a bone. After two half-lives, 11,136 years, how many carbon-14 atoms will there be? *250*
3. Tree rings can tell *All of the above*
4. Carbon-14 is used to date *fossilized organisms.*
5. Potassium-40 is used to get a relative date for *dinosaur fossils.*
6. Half-life measures *the time it takes half the radioactive atoms to turn into stable atoms.*
7. In a column of rock, a possible progression of fossils is *(see illustration on the left)*
8. The order of rock layers indicates *the relative age of the layers.*
9. The Principle of Superposition states that *the oldest layer of rock is at the bottom and the youngest layer is at the top.*
10. Atoms of an element with different numbers of neutrons in the nucleus are called *isotopes.*
11. What isotope of carbon is used for radiometric dating? *Carbon-14*
12. What is the stable isotope carbon-14 turns into? *Nitrogen-14*
13. A layer of igneous rock is running at an angle through layers of sedimentary rock. The igneous rock has an absolute date of 65 million years old. What is the minimum age of the top layer of sedimentary rock through which the igneous rock runs? *More than 65 million years old*

Questions

What is the name of the principle that your answer in question 13? *The Principle of Cross-Cutting Relations*

How do changes in environmental conditions affect selective pressures on organisms? Use polar bears and sea ice as an example to help explain. *Organisms are adapted to certain environments. When the conditions change, they might not have the adaptations they need to survive in the new environment. Polar bears are adapted to hunt from sea ice. If the sea ice melts, polar bears must change eating strategies or they will starve.*

Lesson Review

Evolution: When

Relative Dating = gives a time range for when an organism lived.

1. **Principle of Superposition** = in sedimentary rocks, the oldest layer is at the bottom and the youngest layer is at the top → when fossils are in the rocks → the oldest fossils are at the bottom and the newest ones are at the top

2. **Principle of Cross-Cutting Relations** = a layer of igneous rock is younger than any sedimentary rocks it cuts across → radiometric dating is used to determine an absolute age for the igneous layer → this rock layer gives relative dates for the fossils found in the layers of sedimentary rock it cuts through

3. **Index fossils** = some fossils were so abundant when they lived that they can be used to give a relative date for other fossils found with them

To be a good index fossil:
- The organism must have left a large number of fossils.
- The organism must have lived for a short period of geological time.
- The organism must have been widespread globally.

Radiometric dating = absolute dating = measures the change in amount from a radioactive form of an atom into a stable form

The decay of an unstable isotope of one element into a stable isotope of another happens at a constant rate.

That rate can be used along with the ratio of the unstable isotope to the stable isotope, to give an absolute age for rocks and fossils.

Carbon-14, Uranium-235, and Potassium-40 are unstable isotopes used to date fossils and rocks. The choice of isotopes depends on the age of the fossils or rocks.

Carbon-14 has the shortest half-life – it is used to date fossils 70,000 years old or younger.

Uranium-235 and Potassium-40 have long half-lives; they are used to date igneous rocks near where fossils are found. This gives an absolute age for the rocks but only a relative age for the fossils.

Sedimentary rock cannot be used to date fossils because it is made up of bits of other rocks.

Teacher Guide

Unit V: Evolution

Answer Key Unit Exam Chapters 20–23

The exam for Unit V is found in the appendix of the student Workbook.

1. Multiple Choice. (2 points each = 32 points total)

When the first photosynthetic organisms evolved, they spewed a waste product, *oxygen*, into the air and seas. This waste product in the seas led to the *banded iron formation* that can now be seen in sedimentary rocks around the world. This waste product in the air led to the creation of the *ozone layer*.

When biological evolution occurs, it happens to a(n) *population.*

The definition of population is *a group of organisms of the same species living in the same area.*

The definition of species is *a group of organisms that are genetically related so that they can reproduce, producing offspring that can also reproduce*

A good index fossil *All of the above*

Genetic variation is caused by *mutation and genetic recombination.*

The case of the peppered moth, where dark moths were less likely to be eaten when tree bark darkened and therefore became more common than light-colored moths, is an example of *1 and 3 above*

The process that explains how all species of organisms came to be is called *evolution.*

Some of the strongest evidence for evolution comes from the homologies between different species of organisms. The word *homologies* means what? *Similarities*

The forearms of a dolphin and a lizard are similar. This is an example of *homologous structures.*

Humans have a tailbone at the end of their spine but no tail. This bone is a *vestigial structure* left over from a non-human ancestor in our distant past.

The fossil record shows that *All of the above*

With permineralization, *an organism's soft tissue is replaced with minerals and creates a fossilized organism that is a rock.*

Archaeopteryx fossils are transitional fossils that show the link between *birds and reptiles.*

The Principle of Superposition indicates that *3 and 4*

You have grown up and become a world-famous biologist studying how saber-tooth tigers went extinct. While hiking in Alaska you find a fossil of a saber-tooth tiger that has become exposed by melting glaciers. What radioactive isotope would you use to find out how old it is? *C-14*

2. True or False (2 points for each correct T or F, +2 points for each corrected false answer = 30 points total)

T It is a fact that evolution occurs.

F It is a ~~fact~~ that birds evolved from dinosaurs. *theory*

F A scientific theory is based on the ~~opinion of~~ many scientists. *experiments done by*

T A qwitekutesnute was born with a mutation such that a pocket of skin went from its underarms to its sides. This mutation enabled it to glide short distances between trees. This qwitekutesnute had lots of offspring, many with this trait. This means this trait is heritable.

F A harmful trait is selected ~~for~~ *against* and will lead to ~~more~~ *fewer* offspring with this trait.

T The half-life of a radioactive element is the amount of time it takes for half of the radioactive isotopes to turn into stable atoms.

T__ The more closely related two species are, the more similar their DNA is.

F__ The chemistry of all organisms is ~~not~~ homologous. They are made from the same six elements and the same types of molecules.

T__ Fossils are usually found in sedimentary rock.

F__ The Principle of Cross-Cutting Relations states that a geological feature is younger than anything it cuts across. This means that an igneous layer is ~~older~~ than the sedimentary layers because layers can only cut through a layer if it is already deposited. *younger*

3. Match the events on the timeline (2 points for each correct match = 20 points total)

4.5 – 3.8 billion years ago *the first living organisms*

3 billion years ago *photosynthesis evolves*

2.5 – 2 billion years ago *banded iron formation*

1.85 billion years ago *eukaryotic cells evolve*

1.5 billion years ago *multicellular life evolves*

450 million years ago *plants colonize land*

420 million years ago *animals colonize land*

230 million years ago *dinosaurs roam the earth*

65 million years ago *dinosaurs go extinct*

after 65 million years ago *mammals begin evolving into the many species of today*

4. Vocabulary Match (2 points each = 18 points total)

A. Natural selection | F__an inherited variation that affects the survival of organisms with that trait

B. Genetic drift | C__when a new species evolves from existing species

C. Speciation | B__the random change in the frequency of alleles in a population due to chance events in small populations

D. Overproduction | A__occurs when organisms have a better or a worse chance of survival because of their traits

E. Genetic variation | H__where reproduction is stopped between a population of organisms and other populations within the species

F. Adaptation | I__genetic changes to populations

G. Genetic recombination | D__more organisms are born than the environment can support

H. Reproductive isolation | E__variation of traits between members in a population caused by allelic differences

I. Evolution | G__when homologous chromosomes crossover one another and exchange pieces of DNA

5. Extra Credit. (Answers will vary. Give up to 10 points for a well-thought-out defense of the student's choice.)
A qwitekutesnute is born with opposable thumbs. Choose whether you think this trait is beneficial or harmful.

Example of beneficial: Qwitekutesnutes with this trait are better at grasping food from trees. Because of this they get more food than qwitekutesnutes without opposable thumbs. The increased amount of food makes them healthier, and healthier parents have more babies. More babies mean more qwitekutesnutes with this trait.

Example of harmful: Qwitekutesnutes with this trait cannot climb as well. This makes them more likely to be eaten. Because they are eaten they are not around to have babies. This leads to fewer or no qwitekutesnutes with this trait.

(/100) x 100 =   +   extra credit points

Unit VI: Ecology

Chapter 24: The Biosphere

Introduction to Unit VI

In Unit VI, students will learn about the factors that affect populations and communities of organisms. Chapters 24 and 25 are both on the long side. Students seem to breeze through these chapters, though. This is the unit about life as we see it. It is the area many students think of as "biology." Little did they know, they would have to get through cell biology, genetics, anatomy and physiology, and evolution to get to this point.

The people chosen for the Famous Science Series in Unit VI, Chico Mendez, Jane Goodall, Charles David Keeling, and Rachel Carson, are all people who have had major impacts as defenders of the environment. One student, after completing this unit, made a poster titled "Heroes of the Environment." You might also want to do this with your students.

In the labs for Unit VI, students make a biome diorama, put together a food web based on the wildlife in their own backyard, make a terrarium, and conduct an experiment that looks at the effects of acid rain on plants. The microscope labs go hand in hand with the general lab and theory sections.

Introduction to Chapter 24

Chapter 24 explains how the abiotic, non-living, components of the environment affect organisms. Climate is an abiotic factor that leads to the creation of biomes. A biome is a large area classified by the main plant and animal species native to the area. I divide Earth into five major biomes. Earth can be divided into more biomes than this. For the sake of space, I limited the number of biomes discussed to the five major ones.

In addition, there is an activity included in Chapter 24 where students use information readily available on the Internet to determine the biome where they live. The result may surprise you, especially if you live at elevation.

Note: This is a long chapter to do over two days. I suggest using the three- or five-day schedule.

Learning Goals

- Learn how abiotic factors affect the biosphere.

- Learn six important abiotic factors.

- Recognize that location affects climate.

- Learn that climate determines the location of terrestrial biomes.

- Investigate the characteristics in the one main aquatic and four main terrestrial biomes.

- Learn the names and primary characteristics of the five main biomes.

- Learn about some of the adaptations organisms native to those biomes have.

- Investigate how and why latitude affects the location of each of the three main types of forest.

- Investigate how the quantity of light effects the distribution of life in the aquatic biome.

- Learn about the abiotic factors that determine the biome where you live.

- Research about a famous environmental activist.

- Investigate how different types of soil look different. Think about how that could affect what can grow in the different soil types.

Extracurricular Resources

Books

Exploring Ecosystems with Max Axiom, Biskup, Agnieszka

The Biosphere Realm of Life, Vogt, Gregory

For those of you who live in a city–*Take a City Nature Walk*, Kirkland, Jane

Eyewitness Ecology, Pollock, Steve

Movie

Planet Earth, The Complete Series, starring David Attenborough, this series is a nice complement to watch with Unit 6. Start watching it this week and continue watching it as you work your way through the unit.

Online

Visit Pandia Weblinks for videos and websites recommended for this chapter:

www.pandiapress.com/weblinks-biology2

Lesson

Home, Home on the Biome

This chapter is about the primary abiotic factors affecting organisms and the adaptations they have evolved for dealing with these factors. This chapter explains how abiotic factors affect where organisms live. I present climate as a major abiotic factor. I explain that latitude affects climate, without going into the why of it.

This is a long lesson. It is best done broken up into two days. After the tundra biome is a good place to stop on day one. I have chosen to break the biosphere into four terrestrial biomes and one aquatic biome. The biosphere can be broken into many more biomes. I felt it became overwhelming to include all the different possible biomes. There is a lot of new vocabulary this week. If your students have never studied ecology before, go over it with them to ensure they understand the new terminology. I have scattered the vocabulary throughout the chapter to make it seem like less of a list. Because of that there cases where there are terms that are used and then defined in later paragraphs.

Lab

Biome, Sweet Biome

This is a fun and informative activity. The information is readily available on the Internet for people living in the United States. The biome you live in may not be one of the five main biomes studied in the lesson.

Activity

Diorama

This lab takes students about one hour of planning and three to four hours to complete. It is an art activity, so it could take longer. If students take the time, they will have a diorama that makes them proud. This is a fun project to do together. I suggest choosing a biome and then searching "biome name" diorama on the web, to see what others have done. It is fun to be creative with the material you use. Below is a rainforest biome diorama.

Poetry

A Mad Libs Poem

One student's biome poem:

Home, Home in the Taiga

Home, home in the taiga,
Where the bear and the chipmunks roam.
Where often is seen a coniferous tree,
And the sky has lots of snow clouds all winter.

Home, home in the taiga,
Where animals use a subnivean layer.
Where often is seen a coniferous tree,
And the sky isn't very cloudy in summer.

Home, home on the taiga,
Where wildfires occasionally rage.
Then we get to see baby coniferous trees,
When the fire makes the seeds germinate.

Microscope Lab

Soiled

Choose a variety of soil samples. Different soil samples look quite different when compared to each other. It is interesting to note that the better the soil is for growing, the less interesting it looks with the microscope.

Desert sand

Garden soil

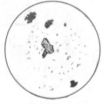

Beach sand

Famous Science Series

Chico Mendes

Note: Chico Mendes was an activist who was trying to stop deforestation in the Amazon Rainforest. He was assassinated by ranchers who wanted to silence him.

What biome was he trying to save? Name the area, country, and continent.
Main biome = Forest, Smaller biome = Amazon rainforest
Brazilian rainforest, Brazil, South America

What was the family business he worked in? How did he go from that to working to save a biome? *Rubber tapping—rubber tapping does not involve killing or cutting down trees. It is a sustainable occupation in this biome, one where the ecological balance is maintained. Mendes was worried rubber tappers would be put out of business because of the forests being cut down by cattle ranchers and people wanting to cut the trees to sell to make furniture.*

In 1987, he flew to Punjab, India, to help stop a project being funded by the Inter-American Development Bank. What was the project he stopped? How do you think this project would have hurt the biome he was trying to save? *He stopped a road being built through the rainforest. The purpose of roads into the forest is to help ranchers and loggers get access to remote and hard or virtually impossible-to-reach locations in the forest, so they can cut down trees. When trees are cut down in the forest biome, it destroys that biome.*

Chico Mendes was assassinated. Why? *Mendes was assassinated by a rancher, because he wanted reserves set up to prevent trees from being cut down. This would have closed those areas to ranchers forever.*

Show What You Know

The Biosphere

1. Write the abiotic factor most important in creating the biome. Circle the adaptations you would expect to find in each of the biomes.

Desert <u>*amount of precipitation is less than 30 cm a year*</u>
burrowing, small size, roots good at absorbing H_2O, specialized kidneys, leaves to prevent water loss, deep root system

Grassland <u>*a wet season that promotes fast growth and a dry season where drought and fires are common*</u>
deciduous trees, burrowing, migration, deep root system

Tundra <u>*latitude and altitude keep temperatures in the tundra low*</u>
asexual reproduction, burrowing, hibernation, migration, small size

Taiga <u>*latitude*</u>
dark green leaves, hibernation, seeds need fire to germinate, subnivean zone, migration, fire-resistant bark, trees grow close together, trees are thin, trees keep leaves all year

Temperate forest <u>*latitude*</u>
deciduous trees, hibernation, subnivean zone, migration, burrows

Rainforest <u>*latitude*</u>

dark green leaves, long arms, prehensile tails, tall trees, few branches until canopy, leaves that shed water

2. Fill in the blank with the best choice from the word box.

Ecologists are scientists who study the *environment*. The entire area of Earth where organisms live is called the *biosphere*. It can be divided into areas called *biomes* that are determined by the *climate* of the area. When these areas are on land, they are called *terrestrial*. When they are in water, they are called *aquatic*. The biotic and abiotic components in an area form an *ecosystem*. Climate is controlled by *abiotic* components, like temperature and precipitation. *Biotic* components are those that are living or once were. All the organisms that live in an area are a *community*. Areas with a lot of different organisms of many species have a high *biodiversity*.

3. Multiple Choice

Earth's tilt creates *uneven heating of the earth*

Periodic fires are common in this biome: *grassland*

Because water absorbs sunlight, the highest biodiversity in the ocean is in the *sunlight* zone where *photosynthesis* takes place.

Most autotrophs in the aqueous biome live in the *sunlight zone.*

The difference between freshwater and ocean water is *the amount of dissolved salt.*

The location of terrestrial biomes is affected by *latitude and location on a continent.*

An ecosystem can be *All of the above*

The forest biome can be divided into *three* groups. The location of each is determined by its *latitude.*

The rainforest covers 6 percent of the earth and *50 percent* of Earth's organisms live there.

Why do deciduous trees drop their leaves? *To conserve energy and water*

Lesson Review

The Biosphere

New vocabulary:

community = all the organisms that live in an area

environment = the area where an organism lives

ecology = the study of the interactions between organisms and their environment

ecologist = a scientist who studies ecology

ecosystem = biotic and abiotic components of an environment

climate = typical weather in an area

biotic = living parts of the environment (or once living – ex. dead leaves)

abiotic = non-living components of the environment

biosphere = entire area of Earth where organisms live

biome = a large area defined by the organisms native to it

biodiversity = <u>number</u> and <u>variety</u> of species of organisms in an area; even if an area has a lot of one species of organism, it has low biodiversity

Abiotic factors:
- tilt of Earth
- temperature, climate and sunlight
- water = precipitation on land
- soil type
- fire

There are four main terrestrial biomes:
1. **Grassland**
   - main abiotic factor = wet season + dry season with drought and fires
   - plant adaptations
     - deep roots
     - fast growth
     - deciduous
     - tree trunks store water and have chemicals to prevent them from catching fire
   - animal adaptations
     - migrate
     - burrow

2. **Tundra**
   - main abiotic factor = latitude and altitude
   - coldest biome, permafrost
   - treeless
   - low biodiversity
   - plant adaptations
     - short
     - asexual reproduction
   - animal adaptations
     - hibernation
     - migrate
     - higher percentage of body fat
     - short extremities
     - burrow

3. **Desert**
   - main abiotic factor = low precipitation
   - low biodiversity
   - plant adaptations
     - roots good at absorbing water
     - leaf shape to prevent water loss
     - lots of seeds
   - animal adaptations
     - small body size
     - specialized kidneys
     - burrow

4. **Forest** – there are three main types of forest biome. Main abiotic factor = latitude
   A. Taiga = 50° – 60° N latitude
   - plant adaptations

     o dark leaves that increase opportunities for photosynthesis

     o leaves stay on trees all year to increase opportunities for photosynthesis

     o shape of tree so snow slides off more easily

     o seeds need fire to germinate

     o thick, fire-resistant bark

     o trees are thin and grow close together

- animal adaptations

     o migrate

     o subnivian zone

     o hibernate

B. Temperate = 23.5° – 50° N and S latitude

- plant adaptations

     o deciduous

- animal adaptations

     o hibernate, migrate, and burrow

     o subnivian zone

C. Tropical = 23.5°N - 23.5°S latitude

- most rain
- highest biodiversity
- half of all species
- covers 6 percent of earth and produces 40 percent of oxygen
- plant adaptations

     o tall trees with few branches until the canopy

     o leaf shape to prevent water runoff, large dark green leaves

- animal adaptations

     o prehensile tail

     o long arms

Aqueous Biome – 2 types = freshwater and ocean

- covers 75 percent of Earth

Abiotic Factors:

freshwater

- rate of water flow

freshwater and ocean

- amount of sunlight
- concentration of dissolved minerals
- temperature of water

ocean

- upwelling, areas of high biodiversity because nutrients are cycled as water comes up from the ocean's depths
- plant and animal adaptations

     o ability to attach to surfaces

- animal adaptations

     o migrate

     o body shape and coloration

Unit VI: Ecology

Chapter 25: Predator and Prey

WEEKLY SCHEDULE

Two Days

Day 1
- ☐ Lesson
- ☐ Lab
- ☐ MSLab

Day 2
- ☐ FSS
- ☐ Lesson Review
- ☐ SWYK

Three Days

Day 1
- ☐ Lesson
- ☐ Lab

Day 2
- ☐ MSLab
- ☐ FSS

Day 3
- ☐ Lesson Review
- ☐ SWYK

Five Days

Day 1
- ☐ Lesson

Day 2
- ☐ Lab

Day 3
- ☐ MSLAB

Day 4
- ☐ FSS

Day 5
- ☐ Lesson Review
- ☐ SWYK

FSS: Famous Science Series
MSLab: Microscope Lab
SWYK: Show What You Know

Introduction

The main topic of Chapter 24 was the abiotic, non-living, factors in the biosphere that affect organisms. The topic of Chapter 25 is the biotic, living, factors that affect organisms.

Learning Goals

- Learn about food webs.

- Understand how food energy is available in lesser amounts going from producers to consumers.

- Learn the main types of biotic interactions: predator/prey, competition, symbiosis

- Recognize the difference between interspecific interactions and intraspecific interactions.

- Learn how organisms adapt to survive.

- Understand why niches are important in minimizing competition.

- Learn the three types of symbiosis.

- Identify the wildlife that lives in your backyard.

- Learn how plant predation affects plants.

Extracurricular Resources

Books

Exploring Ecosystems with Max Axiom, Biskup, Agnieszka

A Journey into Adaptation with Max Axiom, Biskup, Agnieszka

The World of Food Chains with Max Axiom, O'Donnnell, Liam

Totally Amazing Rainforests, Golden Books

Fur, Feathers, and Flippers: How Animals Live Where They Do, Lauber, Patricia

Food Chains and Webs, Wallace, Holly

One of the many books written about or by Jane Goodall; one suggestion is: *My Life with the Chimpanzees*

Online

Visit Pandia Weblinks for videos and websites recommended for this chapter:

www.pandiapress.com/weblinks-biology2

Lesson

No Fighting No Biting. Wanna Bet?

The lesson covers the biotic, living, factors that affect organisms. These interactions create natural selection for specific traits, adaptations. There is selection for morphological traits and of behavioral traits. Predator/prey relationships are a driving force for the selection for traits that help catch prey and elude predators. Competition also drives selection. The strongest competition is between members of the same species, intraspecific competition. Because these organisms have the same needs, they share the same niche. The more parameters of their niche that organisms share (and in the case of organisms of the same species, these parameters are identical), the stiffer the competition between those organisms.

The terms interspecific and intraspecific might cause confusion for students. The prefix *inter* means "among or between." The word *interspecific* means "between species." The prefix *intra* means "within." The word *intraspecific* means "within the same species." These are useful prefixes to know for high school science.

Lab

Backyard Food Web

Set aside enough time so you can relax, enjoy, and spend as much time as possible getting to know all of the organisms in your backyard. It is helpful for students to be able to discuss what they are observing. The wildlife books are optional, but they are very helpful when determining what animals eat. If you know there are animals in your backyard that you do not observe, it is okay to include them. When filling in the lab sheet, make sure it is completed as a food web not as a food chain. Very few animals eat just one type of food.

Make sure students include plants and insects in their food web and help them look for evidence of symbiosis. If you have roses, you might be able to see ants and aphids in action. One type of symbiosis that is definitely going on in your backyard is parasitism. Wild animals have internal worms, fleas, and lice. You can talk about the impact of these parasites to the health of the host, what kind of niche a parasite has compared to other animals (at least they keep them alive as opposed to carnivores), and how and why we keep pets parasite-free.

This is a good time to discuss why there are so many more plants in numbers and in mass than animals. It is because at each level up in the food web, there is less energy available to support the organisms at that level.

Microscope Lab

Plant Predation

Have you ever thought of plants as prey? Probably not, but they are.

Before

After

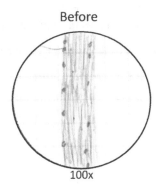

100x

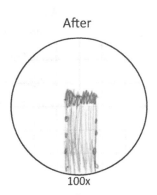

100x

Describe the differences between the two blades of grass.
At the place where the grass was torn, it was much darker than anywhere else that was not torn.

Why do you think grazing can be harmful to the organism to which it happens?
I think it damages the cells at the site where grazing occurs, killing those cells.

Sometimes when grass is cut, the tip turns brown. Use the differences you observed between the two blades to explain why this might occur.
At the tip where grass is torn, the cells die. These dead cells turn brown.

Famous Science Series

Jane Goodall

When researching about Jane Goodall, the problem is the amount of information. Because of the sheer amount of material, it can be hard to sift through it all to answer the questions. She is a prolific author herself. In addition, many people have written about her. You might need to help with the research because of this.

A good title for Jane Goodall could be "Famous Observer of Biotic Interactions." What organisms did Ms. Goodall observe, and why is this a good title? *Goodall observed chimps; answers will vary*

What are the two main threats that chimps face today? *Poaching and loss of habitat*

It has been said that Jane Goodall, through her work, has changed the way we think about all animals. She once said, "Only if you understand, will you care. Animals have feelings too." What did she mean by this? *Answers will vary*

Many of the volunteer and outreach programs started by Goodall reach out to kids your age. How can people your age help save chimps? *Answers will vary*

A famous anthropologist gave Jane Goodall her start. What is his name and why is he famous? *Louis Leakey proved that people evolved in Africa. He also helped push back the date for when hominids evolved by thousands of years.*

The Gombe National Park is the location of the chimps Goodall studied. **What country is Gombe National Park in?** *Tanzania*

What lake does the park border? Find the lake and Olduvai Gorge, where Leakey made his discoveries, on a map of Africa. *Lake Tanganyika*

What famous explorer discovered the lake's only outlet? *Livingstone*

At Gombe National Park, humans need to keep 10 meters away from the chimps. Why? *Chimps and people can transmit diseases to one another.*

Show What You Know

Predator and Prey

1. If you have a pet, look at the list of ingredients on your pet's food. Based on the list, is your pet an omnivore, herbivore, or carnivore? If you do not have a pet, look at a bag of pet food the next time you are at the grocery store. *Answers will vary*

2. Draw a food web using the organisms below. You do not have to use all of them. Draw the arrow going away from an organism to what it might be eaten by. Mark a P on organisms that make their own food.

   Spider ←bee, fly
   Eagle ←snake, cardinal, mouse, trout, chicken (occasionally)
   Lion ←deer, mouse, raccoon, chicken, eagle (if it can catch it)
   Snake ←mouse, cardinal, fish
   Cardinal ←apple tree (P), mushrooms
   Deer ←grass (P), tree (P), mushrooms
   Mouse ← grass (P), tree (P), flower, mushrooms
   Fly ←any of these animals if they are dead (or poop), apple
   Raccoon ←apples from tree, mouse, cardinal, fish, snake, mushroom
   Trout ←fly, bee, spider
   Chicken ←fly, bee, spider, apples from tree
   Person ←deer, chicken, trout, apple from tree, mushroom
   Mushroom ←rotting log, rotting apples

3. List three adaptations predators have for catching prey. *Claws, fangs, poison*

4. List three adaptations prey animals have for avoiding being caught. Explain the benefits of each adaptation.
   - *Camouflage—helps to hide*
   - *Mimicry—look like a bad-tasting or poisonous species so you are not eaten*
   - *Aposematic coloring—says leave me alone, I am poisonous or taste bad*
   - *Bioluminescence helps organisms find food and escape predators*
   - *Plants make bad-tasting or poisonous chemicals, have thorns and stickers, and protect seeds*

5. **Multiple choice**

Cleaning fish will go into the mouth of a barracuda and clean its teeth, eating any parasites they find. This is an example of *All of the above*

Going from producer to herbivore to carnivore: *there is less energy*

When two species have a similar niche, they use *resource partitioning* to reduce competition.

Coral snakes have yellow, red, and black stripes. This is an example of *aposematic coloration.*

Commensalism is a symbiotic relationship where *one species benefits and it doesn't affect the other species.*

The most intense competition is *intraspecific competition.*

Plants defend themselves against predators using *All of the above*

An example of intraspecific competition is *a dog marking its territory.*

The predator/prey relationship is beneficial to a community because *they increase the diversity in the community.*

A population's niche is its *job in the community.*

Questions

6. Plants are called producers because they produce their own food. What is the name of the process they use to do this? In what organelle does this process occur? Write the chemical reaction and state the name of the food made in this process. (10 points if you do not have to peek, 5 if you do)

   Photosynthesis occurs in chloroplasts
   $6CO_2 + 6H_2O + energy\ from\ the\ sun \rightarrow C_6H_{12}O_6 + 6O_2$
   carbon dioxide + water + sunlight → glucose + oxygen

7. All organisms need energy. What is the name of the process used to make energy? In what organelle does this process occur? Write the chemical reaction for this process. (10 points if you do not have to peek, 5 if you do)

   Cellular respiration occurs in mitochondria
   $C_6H_{12}O_6 + 6O_2 \rightarrow 6CO_2 + 6H_2O + energy\ organisms\ can\ use$
   glucose + oxygen → carbon dioxide + water + energy organisms can use

Lesson Review

Predator and Prey

biotic = living parts of the environment
producers = autotrophs = organisms that produce their own food
consumers = heterotrophs = organisms that consume others for food
Types of consumers:
- **carnivores** = eat meat
- **herbivores** = eat plants
- **omnivores** = eat plants and meat

- **decomposers** = use chemicals to break plants and animals into molecules and absorb the molecules

Food web = a diagram showing what organisms eat

Going from producers → herbivores and decomposers → omnivores → carnivores, there is less food energy for each level. Less available food energy leads to smaller population sizes as you go up the energy pyramid.

Natural selection is the process where organisms have a better or a worse chance of survival because of their traits.

Biotic interactions are the interactions between organisms. Biotic interactions lead to natural selection for traits. Selection for traits leads to adaptations that help an organism survive in its environment.

Interspecific interactions = interactions between <u>different</u> species

Intraspecific interactions = interactions between organisms of the <u>same</u> species
Examples of interspecific interactions:

- **Predator/prey relationship**
  - » Predation benefits populations by increasing diversity
  - » There are adaptations to help predators catch prey and to help prey escape predators:
    - Sharp claws and teeth
    - Produce poisonous or bad-tasting chemicals
    - **Camouflage**. Types of camouflage:
      * Disruptive coloration
      * Light bellies for fish and some amphibians
    - Aposematic coloration = warning coloration
    - Mimicry = looks like poisonous or bad-tasting species
    - Bioluminescence = organisms create their own light
    - Stickers or thorns, plants
    - Protective shell
- **Competition**
  - » **Niche** = job an organism has in its environment, such as what it eats and where it lives. Niches reduce interspecific competition.
  - » Intraspecific competition is more intense than interspecific competition, because members of the same population occupy the same niche.
  - » **Resource portioning** = different species of organisms with similar niches use resources slightly differently; this reduces interspecific competition
- **Symbiosis** = interspecific relationship
  - » **Mutualism** = both species benefit from relationship
  - » **Parasitism** = one species benefits from relationship, one species is harmed by relationship
  - » **Commensalism** = one species benefits from relationship, one species is unaffected

Unit VI: Ecology
Chapter 26: Cycles

WEEKLY SCHEDULE

Two Days

Day 1
- ❑ Lesson
- ❑ Lab
- ❑ MSLab

Day 2
- ❑ FSS
- ❑ Lesson Review
- ❑ SWYK

Three Days

Day 1
- ❑ Lesson
- ❑ Lab

Day 2
- ❑ MSLab
- ❑ FSS

Day 3
- ❑ Lesson Review
- ❑ SWYK

Five Days

Day 1
- ❑ Lesson

Day 2
- ❑ Lab

Day 3
- ❑ MSLab

Day 4
- ❑ FSS

Day 5
- ❑ Lesson Review
- ❑ SWYK

Introduction

Chapter 26 looks at the element cycles that are essential to organisms.

Preparation for microscope lab: This lab requires a sample of inoculant. Nurseries and hardware stores sell an inoculant used for coating legume seeds. Soak a small sample of the inoculant in water the night before the lab.

Learning Goals

- Understand elements essential to life cycle through the environment.

- Learn the primary parts of the water cycle.

- Learn the primary parts of the carbon cycle.

- Learn the primary parts of the nitrogen cycle.

- Learn the primary parts of the phosphorus cycle.

- Understand what causes the greenhouse effect.

- Understand how an increase in heat-trapping molecules in the atmosphere is causing global warming.

- Recognize some human behaviors that are contributing to global warming.

- Create a mini ecosystem (a terrarium) to observe the cycling of water, carbon dioxide, and oxygen.

Extracurricular Resources

Books

Carbon-Oxygen and Nitrogen Cycles: Respiration, Photosynthesis, and Decomposition, Harman, Rebecca

The Water Cycle: Evaporation, Condensation & Erosion, Harman, Rebecca

One Well, The Story of Water on Earth, Strauss, Rochelle

Understanding Global Warming with Max Axiom, Biskup, Agnieszka

Camping With the President, Wadsworth, Ginger

Online

Visit Pandia Weblinks for videos and websites recommended for this chapter:

www.pandiapress.com/weblinks-biology2

FSS: Famous Science Series
MSLab: Microscope Lab
SWYK: Show What You Know

Lesson

The Many Cycles of Life

Much of this chapter is taught within the illustrations. Make sure your students do not gloss over these.

Matter cannot be created or destroyed. This means that the elements and molecules on Earth cycle. The first four cycles discussed in Chapter 26 are the four most important cycles for organisms.

The greenhouse effect is a cyclical process as well. It occurs because of heat-trapping molecules in the atmosphere. It is essential to life on Earth. One important heat-trapping molecule is carbon dioxide. The concentration of carbon dioxide molecules is increasing in the atmosphere at an unprecedented rate. The rate of the increase is a problem.

More heat-trapping molecules means hotter average global temperatures. The change is not constant over the globe. The reason the change is not constant has to do with the relationship between the heating and cooling of air in the atmosphere and how this affects precipitation as well as overall weather patterns. It affects life, but the areas of science that explain it are earth science and environmental chemistry. As a result, I do not go into this in detail in this text.

Lab

Cycling Resources

The response to this experiment exceeded my expectations. Students really enjoyed watching water cycle in the terrarium. Even adult observers were curious about why the plants didn't die because of running out of oxygen. I let my students explain that one. A word of warning: Do not overcrowd with plants; they will grow quickly. And do not overwater.

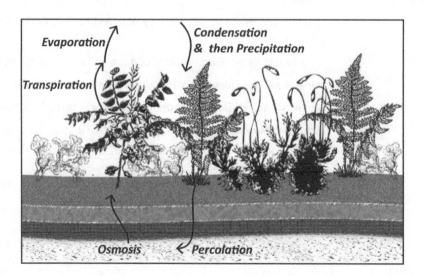

Microscope Lab

Let's Have a Symbiotic Relationship

This microscope lab and the one in Chapter 29 are two of my favorites, along with the first time I looked at an insect's wing with a microscope. When you look at the slide of rhizobacteria with your naked eye, it looks like dirt. But with the microscope you can actually see little brownish organisms moving around. Very cool!

Rhizobacteria can be hard to find. Your best source is the Internet. If it is too costly or hard to find, you might just let this microscope lab go. The microscope lab in Chapter 29 also looks at bacteria with the microscope.

As an alternative plan, you could look at the plants you are using in your terrarium with your microscope. If you choose to do this for your microscope experiment, look at the difference between moss and an angiosperm. Moss does not have vascular tissue. See if you can tell the difference between the two plant types.

What type of symbiotic relationship benefits both species? *Mutualism*

Do you think you will see them fixing nitrogen? *No*

Specimen: *Legume Inoculant* Type of mount: *Wet* Magnification: *400x*

Comments:

There were small pieces that looked like dirt. At first, it was hard to tell if the movement was from water or not. But after 15 minutes, I clearly saw movement of pieces that stopped and started and changed directions. I think these had to be organisms and therefore rhizobacteria.

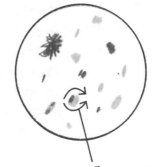

There was circular movement in both directions that looked intentional.

Famous Science Series

Charles David Keeling

What did Charles Keeling begin measuring in 1958? *He began measuring the concentration of carbon dioxide, CO_2, in the air.*

Where did he make these measurements? *Hawaii's Mauna Loa Observatory*

In 1961, data collected by Keeling showed that the amount of carbon dioxide in the air was increasing. When he made a graph from the data it showed a curve that has been named the Keeling Curve. Draw a quick sketch of the Keeling Curve. Remember to label the axes. *The Keeling Curve is a daily record of global atmospheric carbon dioxide concentration maintained by Scripps Institution of Oceanography at University of California, San Diego.*

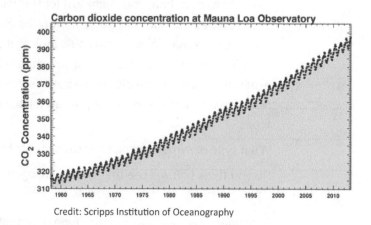

Credit: Scripps Institution of Oceanography

The concentration of carbon dioxide in the air is measured in parts per million (ppm). So, for example, if the measurement for carbon dioxide is 400 ppm, that means that for every one million molecules in the air, 400 of them are carbon dioxide. The zigzagging for each year shows the seasonal fluctuation in concentration. The majority of the world population lives in the Northern Hemisphere. For that reason, during the winter in the Northern Hemisphere, when it is cold, people burn more fossil fuel and the concentration zigs up. During the summer in the Northern Hemisphere, when it is warm, people burn less fossil fuel and the concentration zags down. Overall, however, there has been a steady rise in the concentration of carbon dioxide in the atmosphere.

Based on the Keeling Curve, in the near future, do you think the amount of carbon dioxide molecules in the air will increase or decrease? *Answers may vary, but increase is the most likely response*

Did Keeling show that the increase in carbon dioxide was from artificial sources or from natural sources? *Artificial sources, coming from the burning of fossil fuels (gas, oil, and coal).*

What award did Keeling win in 2002? *The National Medal of Science, the United States' highest award given to a scientist*

Show What You Know

Cycles

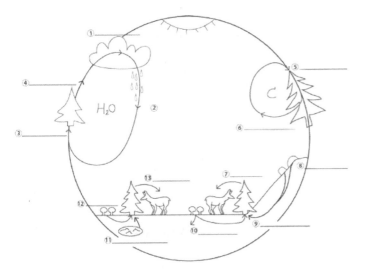

Water cycle
1. _Condensation_
2. _Precipitation_
3. _Osmosis_
4. _Transpiration_

Carbon cycle
5. Photosynthesis
6. _Cellular respiration_

Phosphorus cycle
7. Eating
8. Weathering
9. Uptake by roots
10. Decomposition

Nitrogen cycle
11. Nitrogen fixing
12. Decomposition
13. Eating

Lesson Review

Cycles

Phosphorus, nitrogen, carbon, hydrogen, and oxygen are the elements that make cells (the hydrogen and oxygen come from water).

Carbohydrates = carbon + water
Lipids = carbon + hydrogen + oxygen
Proteins = nitrogen + carbon + hydrogen + oxygen
Nucleic acids = phosphorus, nitrogen, carbon, hydrogen, and oxygen

The chemical reactions in your body happen in water.

The greenhouse effect = the trapping of heat by gas molecules in the environment The concentration of these heat-trapping molecules is increasing in the air because of human activity. This is causing global warming.

Global warming = an increase of the average world temperature resulting from the increased number of heat-trapping molecules in the atmosphere. The problem is the rate of change = how fast the concentration is increasing. The increase is primarily caused by: 1. The burning of fossil fuels, and 2. Deforestation.

Review of genetic unit. Ask your student: DNA has a sugar phosphate backbone. What is one of the elements in phosphate molecules?
Phosphorus (carbon, hydrogen, and oxygen are there too)

Use the circular illustrations in the student textbook to go over the four cycles:

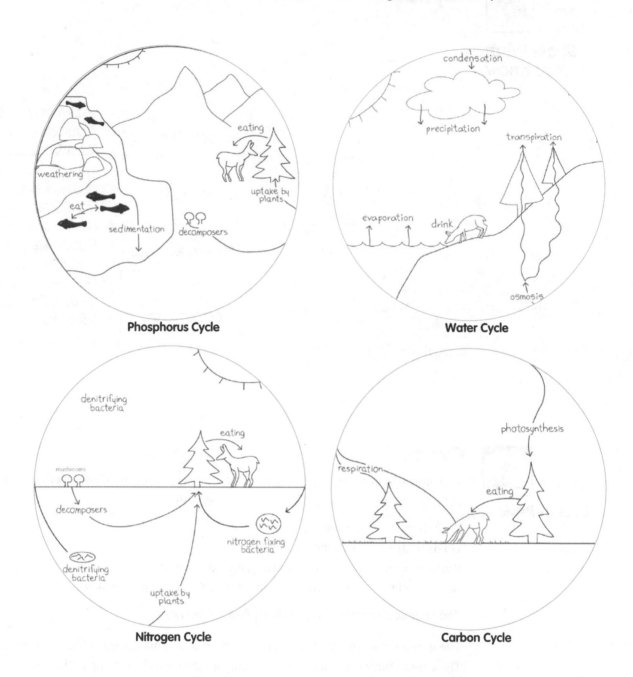

Phosphorus Cycle

Water Cycle

Nitrogen Cycle

Carbon Cycle

Pandia PRESS

Unit VI: Ecology
Chapter 27: Threats

WEEKLY SCHEDULE

Two Days

Day 1
- ❑ Lesson
- ❑ Lab
- ❑ MSLab

Day 2
- ❑ FSS
- ❑ Lesson Review
- ❑ SWYK
- ❑ Unit VI Exam

Three Days

Day 1
- ❑ Lesson
- ❑ Lab

Day 2
- ❑ MSLab
- ❑ FSS
- ❑ Lesson Review
- ❑ SWYK

Day 3
- ❑ Unit VI Exam

Five Days

Day 1
- ❑ Lesson

Day 2
- ❑ Lab

Day 3
- ❑ MSLab
- ❑ FSS

Day 4
- ❑ Lesson Review
- ❑ SWYK

Day 5
- ❑ Unit VI Exam

FSS: Famous Science Series
MSLab: Microscope Lab
SWYK: Show What You Know

Introduction

Chapter 27 examines human impact on the environment. Both the microscope lab and the general lab look at the specific problem of the effects of acid rain on plant life. The Famous Science Series looks deeper into the life of Rachel Carson, the author of *Silent Spring* and the person credited with starting the environmental movement in the United States.

*Possible preparation for How Acids Affect Life Lab: You can grow your own plants or buy plants for this lab. If you grow your own, start three to five weeks before you plan on conducting the experiment. Radish seeds are easy, cheap, and have a short germination time.

This lab takes two to three weeks. There is watering to be done each day until the experiment is completed. This is a good lab to write a formal lab report.

The microscope lab for this chapter takes two days.

This is the last chapter in Unit VI. There is a Unit VI Exam, which covers the material found in chapters 24 through 27, in the appendix of the student Workbook. The answer key is found at the end of this chapter.

Learning Goals

- Learn about the effect global warming has on climate change.

- Understand that the effects of climate change are not the same everywhere and on all types of animals.

- Explain some of the problems caused by pollution.

- Learn how agricultural chemicals that benefit people can pollute and harm other organisms.

- Learn some of the causes of endangering species.

- Learn how loss of habitat negatively impacts organisms.

- Look at a specific example of how habitat is being restored.

- Learn the definition of carbon footprint.

- Examine some things you can do to reduce your carbon footprint and impact on the planet.

Extracurricular Resources

Books

There are so many books on the environment, here are just a few:

A Kid's Guide to Global Warming and Climate Change: How to Take Action, Kaye, Cathryn Berger

From Windmills to Hydrogen Fuel Cells: Discovering Alternative Energy, Morgan, Sally

Sustaining Our Natural Resources, Green, Jen

Recycling, Kallen, Steve

Alternative Energy, Petersen, Christine

Silent Spring, Carson, Rachel

Rachel Carson, Pioneer of Ecology, Kudlinski, Kathleen

Rachel Carson: A Twentieth Century Life, Levine, Ellen

Rachel Carson, Biologist and Writer, Stewart, Melissa

Movies

Happy Feet

Wall-E

Avatar

The 11th Hour

Planet in Peril

A Civil Action

Online

Visit Pandia Weblinks for videos and websites recommended for this chapter:

www.pandiapress.com/weblinks-biology2

Lesson

A Fine Balance

I find the subject matter this week one of the most depressing areas of biology. I have attempted to leave politics aside, and focus only on the science of environmental threats. While politicians might spend a great deal of time arguing about environmental issues as if the problems are negligible, the science doesn't bear them out. I worry about what we are going to do to solve these problems. I think many students feel the same way. The first step is to understand what the problems are. That is what I have attempted to do with this chapter. I believe that people have primarily gotten us into this problem. I also believe it is going to take people to get us out of it. I use the example of the passenger pigeon as an example of an animal that has gone extinct because I want students to see that extinction is not just a problem in the Amazon rainforest. It is a very real problem right in our own backyard. My goal at the end is to make students feel like they can make a difference. People have made a difference in some areas, such as reducing the ozone hole and reducing the incidence of acid rain.

Lab

How Acids Affect Life

*The materials list calls for slightly more vinegar and water than is needed.

The two labs this week both take a look at the effects of acid rain on plants. Acids are reactive with organic (carbon based) material. In the case of many acids, what makes the acid reactive is the concentration of hydrogen ions. A hydrogen atom is a proton and an electron. The most common isotope of hydrogen atoms in their neutral (uncharged) state do not have neutrons. Hydrogen ions are sometimes called protons, because when the electron is stripped away, that is all that is left. Acids have a concentration of hydrogen ions in them. Hydrogen ions are reactive because they strip electrons from other molecules to become neutral. The loss of electrons from molecules can be damaging to cells and therefore to the organisms made from those cells.

Math This Week

Students make solutions for both labs this week. From the standpoint of the experimental result, it wouldn't make much difference if students were slightly imprecise with measurements. But from the standpoint of good science and learning the importance of proper procedure, you should help make sure that students try to measure the volume amounts as carefully and precisely as they can.

Chapter 27: Lab Report

Hypothesis

Water only = I think this will do just fine because I am not watering it with a contaminated solution.

4 water : 1 vinegar = I think it will do better than the plants being watered with more acid but will die eventually, just slower, because the water vinegar solution is the least concentrated.

1 water : 1 vinegar = I think it will do better than the plant being watered with only vinegar, but will die too.

Vinegar only = I think these plants will die the quickest because they are only getting vinegar.

Observations

The plants that received only water stayed basically the same. All the other plants died. The more vinegar that was in the watering solution the faster the plants died.

Results and Calculations

see lab sheets for more details
water only = lived
4 water : 1 vinegar = dead by day 8
1 water : 1 vinegar = dead by day 5
all vinegar = dead by day 4

Conclusions

My hypothesis was correct that plants that were watered with vinegar died. It was incorrect as well. The concentration of vinegar does not make as much difference as I thought it would. It might be that the plants I chose were very susceptible to acid and that other plants would do better and live longer getting the less concentrated acid solution. To learn more I should try this experiment on different types of plants.

Microscope Lab

Acids, Up Close and Personal

This lab looks at how acids affect plants on the microscopic level.

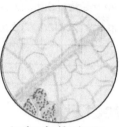 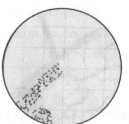

Leaf soaked in vinegar
100x

Leaf soaked in water
100x

Observations: *The leaf in vinegar was brown and had a different texture than the leaf in water. The leaf in water was green and I saw groups of chloroplasts with veins all around. I could not see green chloroplasts in the vinegar leaf.*

Is acid good for leaves? *No*

You already knew water was not harmful to leaves. You included it as a control for this experiment. What does that mean and why is it important to have controls in experiments? *A control is used as a standard of comparison. It helps you determine if the one factor you changed and are testing is responsible for your observations. In the case of this experiment, the only thing you did different for the two leaves was to soak one in vinegar and the other in water. Based on the comparison between the two you can determine that vinegar is harmful to leaves.*

Famous Science Series

Rachel Carson

Most of what I know about Rachel Carson I learned while researching for this chapter. She is a fascinating person. One thing that struck me was her early death. I wonder if her early death from cancer was a result of growing up in such a polluted area, as she did. There would be a lot of irony in that if it was the case. Rachel Carson is a wonderful example of how one person can make a difference.

Carson wrote a famous book entitled *Silent Spring*. **The purpose of the book was to enlighten people about what environmental problem?** *The U.S. Government's spraying of chemical pesticides over fields and forests.*

Carson grew up near a city and two rivers that, at that time, were very polluted. What were the names of the city and rivers, and what was the industry causing the pollution? *Pittsburgh, Pennsylvania. Monongahela and Allegheny rivers. The steel industry*

Carson wrote an article and submitted it to a magazine when she was 10 years old. The magazine published it. What was the name of the magazine? *St. Nicholas*

Carson worked for what agency in the United States government? *The Fish and Wildlife Service*

What was the name of the main pesticide Carson writes about in *Silent Spring*? *DDT*

What U.S. president read Carson's book and created a panel to investigate her claims? *John F. Kennedy*

In 1970, the U.S. government created the EPA. What does this acronym stand for? In 1972, the EPA banned the use of what pesticide in the United States? *Environmental Protection Agency, DDT*

Show What You Know

Threats

Multiple Choice

1. Acid rain *makes soil and water acidic, which kills organisms in the soil and water.*
2. A habitat is *the kind of environment an organism can live in.*
3. Climate change *affects each biome differently.*
4. Pesticides, herbicides, and fertilizers can harm the environment because *they change the chemistry of the environment.*
5. When the water around coral reefs warms, *the photosynthetic algae leaves the coral polyp, and the polyp dies.*
6. The number of animals poached from the rainforest every year is *over 10,000,000.*
7. Salmon ladders are one example of *a device people have created to return habitat back to salmon.*
8. Farming, deforestation, and dams all result in *loss of habitat.*
9. The term *carbon footprint* refers to *the things each person does that produce carbon dioxide.*
10. Littering and the dumping of trash in the ocean has led to *a floating garbage patch that is the size of Texas, off the coast of Hawaii.*

Match the word to the definition.

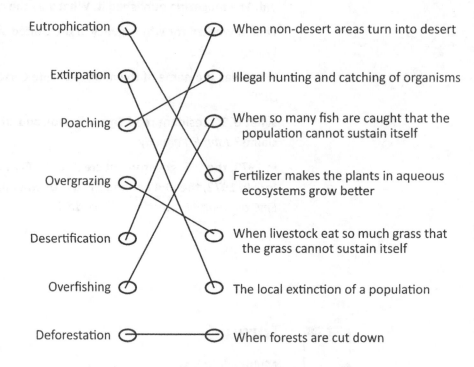

Eutrophication

Extirpation

Poaching

Overgrazing

Desertification

Overfishing

Deforestation

When non-desert areas turn into desert

Illegal hunting and catching of organisms

When so many fish are caught that the population cannot sustain itself

Fertilizer makes the plants in aqueous ecosystems grow better

When livestock eat so much grass that the grass cannot sustain itself

The local extinction of a population

When forests are cut down

List five things you could do to reduce your carbon footprint.
Any five of the following (or any other good ideas your student thinks of):
- *Walk, ride a bike, carpool, or take a bus.*
- *Reduce the amount of trash you generate.*
- *Reduce the amount of electricity you use.*
- *Do not take long showers.*
- *Reuse things*
- *Recycle*
- *Reforestation*
- *Don't litter, don't pollute*

Think of an environmental problem that is affecting the area where you live. Make a plan for what you can do to help. *Answers will vary*

Lesson Review

Threats

Climate Change
- Greenhouse gases are increasing in concentration in the atmosphere.
- These gases trap the heat from the sun.
- The changes are not the same everywhere.
- The problem is the RATE of change.
- Some organisms do fine with the climate changes; some do not. These organisms cannot adapt fast enough to the changes.

Pollution
- Acid rain changes the chemistry of the environment, is bad for organisms
- Oil spills
- Trash
- Agricultural chemicals–benefit people, harm the environment
    - o Herbicides–kill all types of plants not just weeds
    - o Pesticides–are not specific, they can kill all types of animals, even humans in high enough concentration
    - o Fertilizer–increases plant growth, changes the chemistry of the area, benefiting some organisms and harming others.

Endangering Species
- Poaching
- Over-hunting
- Overgrazing
- Overfishing

Loss of Habitat - endangers species
- Deforestation–the habitat is lost to the organisms that lived there
- Dams

What You Can Do
- Reduce your carbon footprint
- Reuse things
- Recycle
- Restore habitat–reforestation
- Don't litter

Unit VI: Ecology
Answer Key Unit Exam Chapters 24–27

The exam for Unit VI is found in the appendix of the student Workbook.

1. Multiple Choice (2 points each, 40 points total)

The most important abiotic factor affecting where different terrestrial biomes develop is *climate.*

Climate change is a problem for organisms because *the climate is changing so fast that organisms do not have time to adapt at the rate of change.*

Tapeworms live in the digestive tract of other animals. When the host animal eats, the tapeworm eats too. The relationship between the tapeworm and the animal it lives in is called *parasitism* and it is a type of *symbiosis.*

The greenhouse effect is most affected by which cycle? *The carbon cycle*

When deer eat grass, there is *less* energy going from grass to the deer.

When a local population of a species goes extinct, it is called *extirpation.*

The uneven heating of the earth leading to differences in temperature based on latitude is caused by *Earth's tilt.*

In a community, different species of animals have evolved different strategies for using the same limited resources; this decreases competition. Scientists say that each population has its own *niche.*

Pollution often occurs when the chemistry of an environment is changed. Common pollutants are *All of the above*

The abiotic factor that has the biggest effect on who lives where in the aquatic zone is *light.*

Global warming is the increase of the average world temperature because of *an increase of heat-trapping molecules such as carbon dioxide in the atmosphere. (Water vapor does trap heat, but it is not the answer here because the amount of it in the atmosphere is not increasing.)*

Overfishing is when so many fish are caught there are not enough to sustain the population. Overfishing is a problem for over *50%* of all species of fish.

The abiotic components in the environment are *those that are the non-living chemical and physical parts.*

Activities that change an area from one where wildlife lives to one where it does not is called *loss of habitat.*

Competition is the weakest between organisms with *the most different* niches.

A tiger's stripes are a type of camouflage called *disruptive coloring.*

The most intense competition is between *the same species of organisms.*

Human causes for the increase of heat-trapping molecules in the atmosphere are *All of the above*

A carbon footprint means the things that *increase* carbon dioxide. It helps the earth when people *reduce* their carbon footprint.

When fertilizers wash into aquatic ecosystems, they make plants grow better. The increase in plant growth uses up oxygen that is dissolved in the water. This decrease in oxygen can kill other organisms. When this happens, it is called *eutrophication.*

2. Food web. Draw a food web on the watery scene below. (2 points each for the nine organisms and correct placement of arrows. 2 points for identifying the producers. 20 points total.) Students should add to the picture to create at least ten organisms that could consume and/or be consumed in this biome. Below is an example. Make arrows from the organism to anything it might be eaten by. Write a P above organisms that are producers.

3. **Biomes.** Title each biome pictured. Then, take one characteristic from each text box and match it with the biome best described by that characteristic. (1 point each, 20 points total)

Aqueous Biome

1. Water everywhere

2. Plants and animals are affected by the amount of sunlight received

3. Animals have sleek, bullet-shaped bodies or can attach themselves to surfaces

Desert Biome

1. Very low amount of precipitation

2. Plant roots are good at absorbing water; leaves are good at preventing water loss

3. Small, burrowing animals with specialized kidneys

Taiga Forest Biome

1. 50°N and 60°N latitude

2. Evergreen trees shaped so snow falls off them without breaking branches

3. Some animals live in a subnivian zone in winter; others hibernate

Temperate Forest Biome

1. Between 23.5° and 50° latitude in both hemispheres

2. Trees are deciduous

3. Some animals live in a subnivian zone in winter; others hibernate

Tropical Rainforest Biome

1. At or near the equator between 23.5°N and 23.5°S latitude

2. Gets the most rain; has the most trees

3. Animals have adaptations for living in trees, such as prehensile tails

Grassland Biome

1. A wet and a dry season; during the dry season fires are common

2. Grasses with scattered trees

3. Grazing herbivores

Tundra Biome

1. Cold temperatures with permafrost

2. Treeless with short plants

3. Migrating or hibernating animals with shorter extremities

4. Fill in the blank. Fill in the essay with words from the text box. (1 point for each correct answer = 20 points total)

The water cycle explains how water cycles through the environment. If we start with a drop of water in a lake, that drop <u>evaporates</u> when the sun warms the water up. This water molecule goes up into the sky, where it <u>condenses</u> and forms a part of a cloud. It doesn't stay there forever, though; this water molecule <u>precipitates</u> back to the earth in a raindrop. Once it hits the ground, it <u>percolates</u> into the ground. Tree roots in the ground absorb the drop using <u>osmosis</u>. This drop doesn't stay in the tree forever; soon enough it evaporates from the leaves by <u>transpiration</u> and it is back up to the clouds for the little drop.

During the water cycle, a water molecule is always in the same molecular form, with two hydrogen atoms bonded to one oxygen atom. During the nitrogen cycle, nitrogen changes the molecules it is in, but it still remains nitrogen. Nitrogen gets into the air by <u>denitrifying bacteria</u> in the soil that release nitrogen into the air. The nitrogen floats around until <u>nitrogen-fixing bacteria</u> change it into nitrogen that organisms can use. In the soil there are also consumers called <u>decomposers</u>; they break down molecules in dead organisms and release nitrogen into the soil. The nitrogen <u>diffuses</u> into plants or is <u>eaten</u> by animals, which make <u>DNA</u> and <u>proteins</u> with it.

Phosphorus is also needed to make <u>DNA</u>. Phosphorus comes from the <u>weathering</u> of rocks. Some organisms get their phosphorus from what is dissolved in the water; others get it by <u>eating</u> it or taking it up through roots with <u>diffusion</u>.

The molecules that make organisms have a lot of carbon in them. In addition, it is the carbon cycle that feeds life on Earth. Plants make their own food with <u>photosynthesis</u>. Animals get this food by <u>eating</u> plants. Both plants and animals get the energy they need with <u>respiration</u>.

Point total: */100*

Unit VII: Classification

Chapter 28: Taxonomy

WEEKLY SCHEDULE

Two Days

Day 1
❑ Lesson
❑ Lab
❑ FSS

Day 2
❑ Research
❑ Lesson Review
❑ SWYK

Three Days

Day 1
❑ Lesson
❑ Lab

Day 2
❑ FSS
❑ Research

Day 3
❑ Lesson Review
❑ SWYK

Five Days

Day 1
❑ Lesson

Day 2
❑ Lab

Day 3
❑ FSS

Day 4
❑ Research

Day 5
❑ Lesson Review
❑ SWYK

FSS: Famous Science Series
MSLab: Microscope Lab
SWYK: Show What You Know

Introduction Unit VII

Unit VII covers systems of classification. If the only system you are familiar with is the old Linnaean system (Kingdom, Phylum, Genus...), you are not alone. There has been a recent upheaval in this field, though, and this system is now primarily used for naming organisms.

The new system used by scientists is called cladistics. Cladistics looks at the evolutionary relationship between organisms, which is determined through genetics. What makes one species different from another are differences in their genomes. It makes sense that the main method used to classify organisms would depend on these genomes. It was not until recently that it was possible to compare the genomes of organisms to see how they relate; that is why this field has grown so much in recent years and will continue to grow.

Classification is covered in five chapters:

Chapter 28 - Explains taxonomy, the branch of biology that classifies organisms, and binomial nomenclature, the system used for naming them.

Chapter 29 - Explains the classification of the prokaryotic organisms that make up domains Bacteria and Archaea.

Chapter 30 - Explains the classification of the eukaryotic organisms classified as plants.

Chapter 31 - Explains the classification of the eukaryotic organisms classified as animals.

Chapter 32 - Explains the classification of the eukaryotic organisms classified as fungi and protists.

In addition to the normal workload for this unit, there is a research report. The essay portion of the report is optional. You are the teacher. You know your time constraints. If writing is the issue, your student could research the topic and relate their findings to you orally.

Introduction Chapter 28

This chapter covers two of the systems of classification. It instructs students how to name organisms and how to fill out a cladogram using a list of organisms and derived traits. Students will get continued practice naming organisms, and working with cladograms in Show What You Know throughout the rest of the course.

Learning Goals

- Learn about the science of taxonomy.

- Understand how and why the science of taxonomy has changed.

- Learn how organisms are named using binomial nomenclature.

- Learn the eight taxonomic levels of classification for organisms.

- Learn about the system of classification called cladistics.

- Learn how to fill in a cladogram.

- Become familiar with the term phylogeny.

- Learn how to fill out and make a dichotomous key.

Math This Week

Dichotomous keys and cladograms both rely on a logical analysis to help classify an organism. Both rely on you to make a series of choices to distinguish between different types of organisms.

Extracurricular Resources

Book

Carl Linnaeus: Father of Classification, Great Minds of Science, Anderson, Margaret J.

Online

Visit Pandia Weblinks for videos and websites recommended for this chapter:

www.pandiapress.com/weblinks-biology1

Lesson

A Rose by Any Other Name . . .

The field of taxonomy has burgeoned in the last two to three decades, because of more sophisticated equipment and testing procedures. As a result, there are now several competing systems of classification. I presented the two most commonly used. I have presented the two theories as if they are complementary, which most scientists do not think they are. One of the problems is that the Linnaean system of taxonomy assumes a hierarchical structure. This misconstrues how modern biologists think of the tree of life. Some taxonomists want to get away from using the Linnaean system completely. The problem with other systems of classification, cladistics in particular, is that it does not have a good method for naming organisms in a clear way to show their relationship. There is a movement to incorporate Linnaeus' binomial naming system without including the rest. For now, I felt both had to be taught. In grades below the college level, the Linnaean system is often the only system taught or tested. In college, cladistics is the most prominent system taught. I wanted students to be prepared for both situations.

It is important that students use italics or underlining for scientific names. Otherwise it is considered only partially correct.

What is the scientific name for a coyote? *Canus latrans*

What is the scientific name for a gray wolf? *Canus lupis*

Lab

Dichotomous Key Mystery

If your students are like mine, they have enjoyed the qwitekutesnutes. A dichotomous key with them should be no different. The purpose of the first part of the lab is to familiarize kids with how to use a dichotomous key. Help your student when they make their own dichotomous key as minimally as possible, so that you can be used as a solver. However, some students need more help. If your student is stuck, go back and talk through how a dichotomous key is constructed by choosing four qwitekutesnutes and talking about the differences between them and how those could be used to segregate each into its own category.

Zoo Destinations

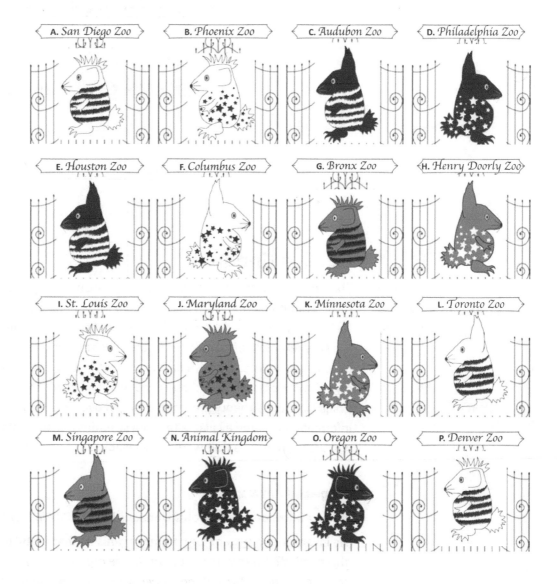

My Leaf Dichotomous Key

Below is an example of one way a dichotomous key can be created for five leaves. In this example, there is a place for each leaf starting at the second juncture, rough texture/smooth texture. Also some paths—central vein, rough texture, etc.—end early in the key.

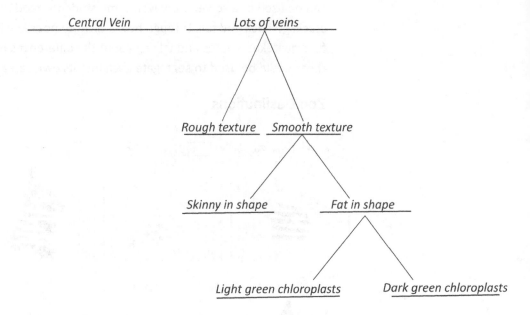

Famous Science Series

Carolus Linnaeus

The Linnaean system of classification may be out of vogue, but he is certainly the most famous taxonomist ever. His system for naming organisms will be around for a long time even if the rest of his system becomes obsolete. Linnaeus' theory is a great example of how scientific theories grow and change, sometimes even becoming obsolete, but with part of them still used, like his system for naming organisms. He began with three kingdoms, one of which, minerals, was not even living. Linnaeus's theory grew and changed over the decades and centuries as more information became available through scientific experimentation from equipment and testing methods that were not available during Linnaeus's time.

Suggested Answers

When and where was Carolus Linnaeus born? *1707 in Stenbrohult, Sweden*

He invented the system for naming organisms that is still used today. Why was this needed? *There was no standard. One organism might have several different scientific names. Many of the names were long with many words. He streamlined that to two words: the binomial system.*

What name did he give humans? What does that name mean? *Homo sapiens, "wise man"*

Linnaeus proposed a system of classification with three kingdoms. What were they? *animal kingdom, plant kingdom, mineral kingdom*

Why don't you think he proposed a kingdom for bacteria? *Anton Van Leeuwenhoek did see bacteria in the 1600s, but it was not until two centuries later that scientists began studying bacteria.*

Who proposed the five-system kingdom? When was it proposed? What were the names of the five kingdoms? *Robert Whittaker, 1969, 1.) Animalia 2.) Plantae 3.) Fungi 4.) Protista 5.) Monera*

One of the kingdoms was split into two kingdoms. Which kingdom was it and why was it changed? *The kingdom Monera was changed to Archaea and Bacteria, when the evidence showed how different the two types of prokaryotes are from each other.*

Discussion question: The system of classification proposed by Linnaeus changed over time. It is a good example of how a scientific theory can grow and change as new information becomes available. Explain what is meant by that statement. *Answers will vary*

Research

Species Spotlight

This assignment is for a research report. It includes researching a living species of the student's choice, completing the Species Spotlight Research Report form, creating a bibliography, and writing a five-paragraph essay. Ideally your student will complete all parts of the assignment and attach them into one report with a cover page. At minimum, your student should complete the Species Spotlight Research Report form and the bibliography. The essay is optional. The essay can be either a factual report or a fictional story with factual elements. The essay is a highly recommended assignment that can be started during this chapter and completed before the end of the course. Time constraints and your student's writing and research abilities will impact your decision to assign the essay. You could choose to modify the assignment into an oral report or software presentation.

If your student has never written a report or essay before, this would be an excellent opportunity to teach these valuable skills. Not only will the process of writing and research teach him much about a particular species, it is also a tool for learning organization of ideas, paraphrasing information, and proper citation of works—all skills that can be applied to many subjects and classes throughout his school career.

In the appendices of your student's Workbook is information regarding proper attribution of sources, plagiarism, and creating a bibliography. These recommendations are based on MLA standards for writing a research paper and attributing a source. You can use this format, or use another of your choosing.

Show What You Know

Cladogram Tutorial

1. Go down the list of traits, asking which trait is shared by all organisms. In this case it is multicellular. Write that on the bottom horizontal line.

2. Ask which organism is multicellular but not a heterotroph, does not have feathers, and cannot fly. That would be the palm tree. Write "palm tree" at the top of the lowest vertical line.

3. Ask which of the three remaining traits is shared by the three remaining organisms. In this case it is heterotroph. Write that on the second to the bottom horizontal line.

4. Ask which organism is a multicellular heterotroph but does not have feathers and cannot fly. That would be the shark. Write shark at the top of the 2nd lowest vertical line.

5. Ask which of the two remaining traits is shared by the two remaining organisms. In this case it is feathers. Write that on the next horizontal line.

6. Ask which organism is a multicellular, heterotroph (with feathers) that cannot fly. That would be the penguin. Write "penguin" at the top of the next vertical line.

7. Fill in the last trait and the last organism at the top of each space. As a check make sure that organism has each of those traits. Ask, can the owl fly, does it have feathers, is it a multicellular heterotroph? The answer is yes.

Taxonomy

List the eight levels of classification in order. Start with the level that has the most species in it and ending with the level that has the fewest species in it.

1. *Domain*
2. *Kingdom*
3. *Phylum*
4. *Class*
5. *Order*
6. *Family*
7. *Genus*
8. *Species*

Multiple Choice

1. The three domains are *Archaea, Bacteria, Eukarya.*

2. Shared traits are *traits that were inherited from a common ancestor.*

3. The branch of biology that classifies and names organisms is *Taxonomy.*

4. The naming system Linnaeus developed for naming species is called *Binomial nomenclature.*

5. Modern taxonomy uses *All of the above*

6. Which is the correct way to write the scientific name for a koala bear? <u>Phascolararctos</u> <u>cinereus</u>

Use the diagram to answer questions 7–10.

7. The name of the diagram is a *cladogram.*

8. The rabbit is most closely related to the *rat.*

9. Which animal has all the shared derived traits listed? *Rabbit*

10. Which shared trait is shared by all animals? *Fur*

Cladograms

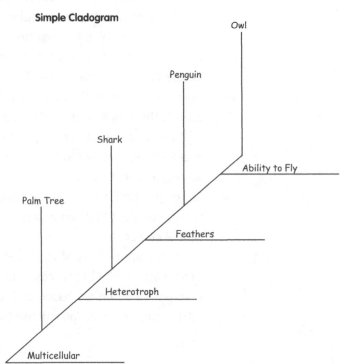

Basic Cladogram

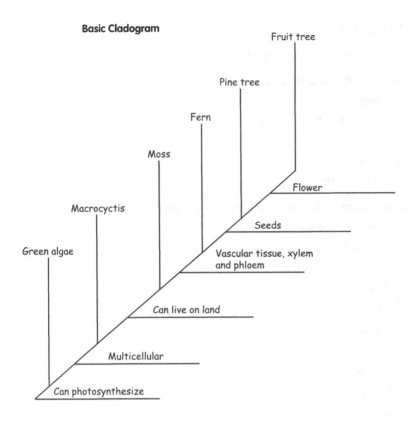

Advanced Cladogram

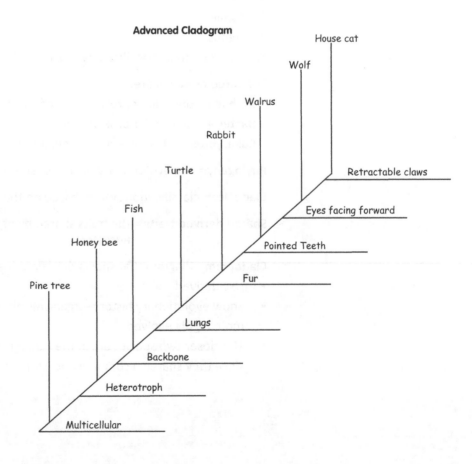

Lesson Review

Taxonomy

Taxonomy = science of naming and classifying organisms

Organisms are classified by:
- Appearance
- Method of getting nutrients
- Anatomy—external and internal
- Genetics
- Cell structure

Binomial nomenclature = how to name organisms
- Name has two words
  - Genus is uppercase
  - Species is lowercase
- In Latin
- Name is in italics or underlined

There are eight taxonomic levels:
1. Domain—most general, most types of organisms
2. Kingdom
3. Phylum
4. Class
5. Order
6. Family
7. Genus
8. Species—most specific, only one type of organisms

The three domains are:

Archaea = unicellular prokaryotes, not bacteria

Bacteria = unicellular prokaryotes

Eukaryotes = all organisms with eukaryotic cells

Phylogeny = the evolutionary history of species

Cladistics = classification system based on the evolutionary history of species

Shared derived traits = the traits shared by organisms who have a common ancestor

Cladogram - diagram that shows the relatedness of the organisms
- Use derived traits
- Show evolutionary history—organisms higher up in the cladogram are not more highly evolved
- The closer two species are in the cladogram, the shorter the amount of time since they shared a common ancestor.

Unit VII: Classification
Chapter 29: Domains Bacteria and Archaea

FSS: Famous Science Series
MSLab: Microscope Lab
SWYK: Show What You Know

Introduction

Chapters 29 through 32 look at the characteristics used to put organisms into one of the six kingdoms. Chapter 29 identifies the characteristics used to place organisms in domains Archaea and Bacteria. Both of these domains have only one kingdom under them, so for organisms in these categories, the domain and kingdom ranking are equivalent. You might have learned about archaea and bacteria as fitting into kingdom Monera. This is an outdated kingdom name. As a result, I do not teach about this in this course.

There is only one lab this week, a microscope lab titled "The Good Guys." It is a nice lab for a formal lab write-up. This lab can be done in one day if you take a sample of yogurt out the day before performing the lab. This makes it easier to compare the amounts of bacteria between the samples. If you do this, be sure and keep track of which slide comes from which sample.

Learning Goals

- Recognize the shared traits in domains Archaea and Bacteria.

- Learn the traits that differentiate the organisms in domains Archaea and Bacteria.

- Learn some common structural features of a bacterium.

- Learn the features used to classify different species of bacteria.

- Learn about organisms in domain Archaea and how they are classified.

- Isolate, identify, and observe different types of bacteria in yogurt.

Extracurricular Resources

Books

The Surprising World of Bacteria with Max Axiom, Biskup, Agnieszka

The Benefits of Bacteria, Sneddon, Robert

Bacteria: Staph, Strep, Clostridium, and Other Bacteria, Wearing, Judy

Archaea: Salt-Lovers, Methane-Makers, Thermophiles, and Other Archaeans, Barker, David M.

This book includes organisms that are not bacteria and archaea: *Human Wildlife: The Life That Lives on Us*, Buckman, Robert

Online

Visit Pandia Weblinks for videos and websites recommended for this chapter:
  www.pandiapress.com/weblinks-biology2

Lesson

How Could They Be So Different?

The organisms in domains Bacteria and Archaea are as different from each other as they are from those in domain Eukarya. They do both have only unicellular organisms in them and they both have only prokaryotic organisms in them. The differences are primarily genetic and biochemical. The biochemical steps that go into making the molecules that make cells, organelles, and molecules like RNA are conservative. That means they don't change much. When organisms use completely different types of material to build cell membranes, for example, that is a BIG difference. I do not go into the details about the differences, frankly, because students do not have the chemistry knowledge to do so.

There is much more information about bacteria than there is about archaea. This reflects the state of knowledge surrounding the two domains.

Microscope Lab

The Good Guys

This lab takes two days with yogurt sitting at room temperature overnight between Day 1 and Day 2. You could complete the lab in one day if you divide the yogurt into two parts; keep one part refrigerated, and place the other on the counter 24 hours before the lab.

Bacteria are smaller than eukaryotic cells; even so, you can see them easily at 100x and 400x magnification. Be careful not to get yogurt that has fillers added like fruit or sugar or heaven forbid, M&M's. Make sure, also, that you get yogurt with live active culture; this is the good bacteria your students will be looking at.

Three of the types of bacteria in the samples I looked at were rod-shaped. I could see slight differences in them, but there was no obvious way to tell them apart. The streptococcus looked like little balls chained together; they were easy to see.

You might be surprised to see the bacteria stop moving after you add stain. Adding stain kills the bacteria. I would suggest staining, though; there is better definition of the organisms with the stain.

Suggested Answers

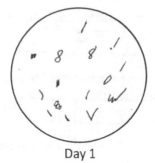

Day 1

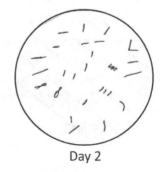

Day 2

Pandia PRESS

Day 1

Read the label of the yogurt container. Write the names of the types of bacteria in the list of ingredients.

Lactobacillus acidophilus

Lacto bulgaricus

Streptococcus thermophilus

Bifidus = Bifidobacterium

Record the types of bacteria you see with your microscope. Which bacteria are most abundant? Which bacteria are least abundant?

Lactobacillus acidophilus—rod-shaped

Lacto bulgaricus—rod-shaped

Streptococcus thermophilus—spheres connected as a chain

Bifidus = Bifidobacterium—rod-shaped

I saw lots of rod-shaped bacteria; some looked different from others and some of the rods seemed linked at their ends. There were several chains of spheres. I saw as few as two spheres linked and as many as six linked together. The rods were more abundant than the spheres.

Day 2

Did you see anything new? Did the abundance of any type of bacteria change more than others? Give explanations with your answers. *I didn't see anything new, but there were about twice as many bacteria. I think the bacteria increased in number overnight using asexual reproduction.*

We have a symbiotic relationship with the good bacteria found in yogurt. These bacteria help us digest food and fight pathogens that get inside us. What benefit do we provide to the bacteria? *We provide a nice, safe, warm place to live with plenty of food and water.*

Famous Science Series

The Archaea of Yellowstone

The examples of archaea used in this book are all extremophiles, organisms that live in extreme environments. Some people erroneously think all archaea are extremophiles, because the first archaea discovered were from environments originally thought to be too inhospitable to support life. Since that time, archaea have been found in environments that are within the parameters of "normal." Archaea are everywhere, just like bacteria. They even live deep in the oceans at hydrothermal vents. Archaea are so small it can be hard to find them.

Research archaea found in Yellowstone National Park. What colors are the mats of archaea?

Yellow, orange, green, brown

What is the temperature of the water leaving the ground at the geysers in Yellowstone? *204°F, 95.5°C*

Do archaea live in the hot water around these geysers? *Yes*

At what temperature does water boil? *212°F, 100°C*

The American Burn Association recommends your bathwater be what temperature? *100°F, 37 °C*

Research information about the temperatures at the thermal pools in Yellowstone National Park. Could you live in the water of these pools? *No, the water in all the pools is above 100°F, 37°C*

Show What You Know

Domains Bacteria and Archaea

Questions

1. This is the classification for the pathogenic bacteria that is the leading cause of bacterial pneumonia.

Domain	*Bacteria*
Kingdom	Bacteria
Phylum	Firmicutes
Class	Bacilli
Order	Lactobacillales
Family	Streptococcaceae
Genus	*Streptococcus*
Species	*pneumoniae*

2. What shape do you think streptococcus bacteria have? Do they occur in groups of two, clusters, or in chains?

   Strepto = chains and coccus = spherical

   Streptococcus = spherical chains

3. All cells have ribosomes, whether they are bacteria, archaea, or eukarya. Why? Hint: What do ribosomes do that is so important? *Because all organisms need proteins, and ribosomes make proteins.*

Multiple Choice

1. Why was the new classification of archaea made? *Archaea use very different materials to build cellular structures.*
2. Bacteria are responsible for *All of the above*
3. What shape are lactobacillus bacteria? *Rod-shaped*
4. A pathogen is *disease-causing.*
5. Which of the choices is NOT a trait of bacteria? *They are eukaryotic.*
6. Bacteria are classified by *All of the above*
7. Bacteria use their flagellum to *move.*
8. The DNA of archaea and bacteria is *in a circular ring.*
9. Archaea are classified using *differences in their RNA.*
10. Cyanobacteria are autotrophs because *they have chlorophyll in their cell membranes, which they use for photosynthesis.*
11. Which of these is NOT true of archaea? *They have RNA but no DNA.*
12. Anaerobic organisms *cannot survive in the presence of atmospheric oxygen.*
13. You know from the name that spirillum bacteria are *spiral.*
14. Pili help bacteria *attach to surfaces.*
15. Archaea are sometimes called extremophiles because they *live in extreme environments that would kill other organisms.*

Lesson Review

Domains Bacteria and Archaea

There are two main types of cells:
1. Prokaryotic = no nucleus
2. Eukaryotic = nucleus

All prokaryotes are in domains Archaea and Bacteria

Organisms in domains Archaea and Bacteria
1. Unicellular
2. Prokaryotic
3. Microscopic
4. DNA is a double strand in a ring shape
5. Reproduce asexually
6. Have the three things ALL cells have = cell membrane, cytoplasm, DNA

The differences between organisms in domains Archaea and Bacteria
1. They use different material to build
   - ribosomes
   - cell walls
   - cell membranes
2. RNA

3. Genetics
4. Biochemical differences

Bacteria are classified by
1. Shape and how they group
2. Feeding strategy
3. Conditions they grow in
4. Similarities and differences in DNA

Review parts to a bacterium cell:

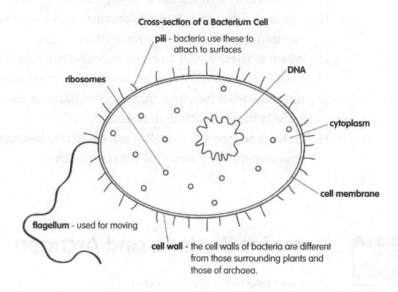

Cross-section of a Bacterium Cell

pili - bacteria use these to attach to surfaces

ribosomes

DNA

cytoplasm

cell membrane

flagellum - used for moving

cell wall - the cell walls of bacteria are different from those surrounding plants and those of archaea.

Archaea are classified based on differences in their RNA.

Unit VII: Classification
Chapter 30: Kingdom Plantae

WEEKLY SCHEDULE

Two Days

Day 1
- ❑ Lesson
- ❑ Lab
- ❑ MSLab

Day 2
- ❑ FSS
- ❑ Lesson Review
- ❑ SWYK

Three Days

Day 1
- ❑ Lesson
- ❑ Lab

Day 2
- ❑ MSLab
- ❑ FSS

Day 3
- ❑ Lesson Review
- ❑ SWYK

Five Days

Day 1
- ❑ Lesson

Day 2
- ❑ Lab

Day 3
- ❑ MSLab

Day 4
- ❑ FSS

Day 5
- ❑ Lesson Review
- ❑ SWYK

Introduction

Domain Eukarya is also called domain Eukaryota. As the name indicates, the organisms in domain Eukarya all have eukaryotic cells. There are four kingdoms in domain Eukarya: Plantae, Animalia, Fungi, and Protista. This chapter focuses on kingdom Plantae.

Learning Goals

- Review that ALL organisms in domain Eukarya have eukaryotic cells.

- Review the names of the kingdoms in domain Eukarya.

- Learn that taxonomic classification uses division instead of phylum for plants.

- Learn the traits ALL plants share.

- Become familiar with the branched bush diagram using the four main divisions for plants.

- Learn the primary traits used for the placement of plants in their division.

- Learn the traits of non-vascular spore producing plants in division Bryophyta.

- Learn the traits of vascular spore producing plants in division Pterophyta.

- Learn the three main advantages of seeds versus spores.

- Learn the names of the three types of gymnosperms and the traits of all plants in division Gymnospermae.

- Learn the traits of plants in division Angiospermae.

- Learn the traits of the two main classes within division Angiospermae: Monocotyledon and Dicotyledon.

- Learn about the plant life in your backyard.

Extracurricular Resources

Books

George Washington Carver Botanist, Adair, Gene

Shanleya's Quest A Botany Adventure for Kids Ages 9 to 99, Elpel, Thomas J. This is an interesting story approach to plant identification. It is the story of a girl who takes her canoe island-hopping. On each island she looks at the traits and then the book identifies the plant based on the traits.

Plants & Fungi Multicelled Life, Snedden, Robert

Online

Visit Pandia Weblinks for videos and websites recommended for this chapter:

www.pandiapress.com/weblinks-biology2

Lesson

Kings of Their Domain

Chapter 12 and Chapter 13 can be used to review some of the terms and concepts in this chapter.

There are more than four divisions of plants. These are the four main divisions with at least one representative for each branch of the bush.

Chapter 29 and Chapter 30 both have what is called a branched bush diagram at the start. The purpose of the branched bush diagram is to show the four kingdoms in domain Eukarya. They are in the order in the diagram that they are discussed in the book.

Does it feel like Chapter 30 is one list of traits after another? Traits are what are used to classify organisms.

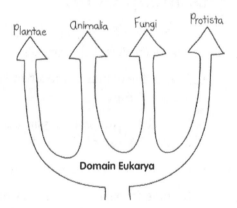

Lab

Perusing Plants

A field guide can add a lot to this lab. If you do use a field guide, try to find one that has local plants in it.

I recommend teacher participation with this lab. Help students initially when they are examining different plants to determine the division. Try to find plants in all four divisions.

Microscope Lab

Distinguishing Dicots and Monocots

The two classes of leaf look dramatically different under the microscope. Make sure your students get a leaf from both a monocot and a dicot. A blade of grass is a good source for a monocot leaf. When locating a tree that is a dicot, make sure the tree is an angiosperm. Make sure the tree is not a bamboo tree or a palm tree.

As an addition to this lab, your students can take a sample of moss and make a wet mount slide and compare it to the monocot and dicot leaves, to see if they can tell the difference between a plant with and without vascular tissue.

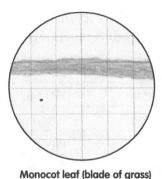

Monocot leaf (blade of grass)

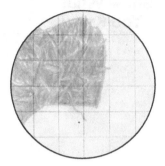

Dicot leaf (tree leaf)

Famous Science Series

George Washington Carver

I recommend reading *George Washington Carver Botanist* by Gene Adair, or another biography about Carver. His life story is compelling.

When and where was George Washington Carver born? What was the situation of his birth? *He was born in Diamond, Missouri in either 1864 or 1865. He was a slave.*

Carver received a bachelor's degree in agriculture in 1894. Where was it from? *Iowa State University*

While he was studying for his master's, Carver was approached by another famous African-American man about teaching at the college where he was principal. Who was the man and what was the name of the college? *Booker T. Washington, Tuskegee Normal and Industrial Institute*

Carver was a researcher as well as a teacher. He studied plants. What plants did he study? *Peanuts, soybeans, pecans, and sweet potatoes*

Carver invented 300 uses for one of these. What was it? *Peanuts*

George Washington Carver developed crop-rotation methods that helped farmers in the South. What is crop rotation and how does it help farmers? What two crops did Carver rotate and why did this help? *Crop rotation is the practice of growing crops of different plants in a location, one after the other. If done correctly, it helps to enrich the soil. At the time Carver was working on crop rotation the main crop grown in the South was cotton. Cotton depletes soil of its nutrients. The crop-rotation method developed by Carver was to plant cotton one year and the next year plant peas or beans, plants that have a symbiotic relationship with rhizobacterium, which adds nitrogen back to the soil. This greatly increased the quality of the cotton and the soil.*

Pandia PRESS

Show What You Know

Kingdom Plantae

Questions

1. Write the classification for a species of bristlecone pine in the chart below. Write the words outside the box in the correct places in the table.

Domain	Eukarya
Kingdom	Plantae
Division	Pinophyta
Class	Pinopsida
Order	Pinales
Family	Pinaceae
Genus	*Pinus*
Species	*longaeva*

What is the scientific name for this species of bristlecone pine? *Pinus longaeva*

2. Fill in the cladogram. Put the shared derived traits on the horizontal lines and the organisms at the top of the vertical lines.

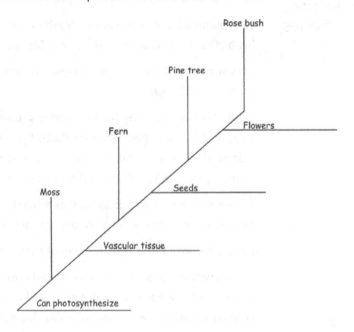

3. Each one of the statements about plants is false. Fix each statement to make it true.

All plants...
- are ~~unicellular~~. *multicellular*
- are ~~mobile~~. *immobile*

- have ~~prokaryotic~~ cells. *eukaryotic*
- have cell walls made from ~~glucose~~. *cellulose*
- ~~are heterotrophs, that use cellular respiration to make food~~. *are autotrophs, which use photosynthesis to make food.*
- reproduce using ~~binary fission~~. *spores or seeds*

Multiple Choice

1. You are walking along a creek and you see a clump of low-growing plants. When you examine the clump more closely, you see it has no roots or leaves. It is a *moss.*
2. You walk further and see a plant with an interesting leaf. You turn the leaf over and see rows of button-like spores running down it. It is a *fern.*
3. Next, you reach down and pick a flower. It is a(n) *angiosperm.*
4. The flower has 9 petals. It is a *monocot.*
5. Another name for a bryophyte is a *moss.*
6. Another name for a pterophyte is a *fern.*
7. Seeds *All of the above*
8. A plant with no vascular tissue is a(n) *bryophyte.*
9. A plant with vascular tissue that uses spores to reproduce is a(n) *pterophyte.*
10. A plant that uses seeds arranged on cones to reproduce is a(n) *gymnosperm.*
11. A plant that makes flowers is a(n) *angiosperm.*
12. A plant's vascular system *transports materials to all parts of the plant.*
13. If a daisy has 88 petals, it is a *dicot.*
14. Bamboo has long, parallel veins in its leaves. It is a *monocot.*
15. Spores *need water for part of their life cycle.*

Lesson Review

Kingdom Plantae

Try using the Socratic approach and see how well students remember the traits with or without a little prompting.

Domain Eukarya has four kingdoms:
1. Plantae
2. Animalia
3. Fungi
4. Protista

* Plant vascular tissue = xylem + phloem

* Plant vascular tissue transports food, water, and nutrients throughout the plant

* Seeds and spores are both capsules containing embryonic plants that will grow into adults

* The taxonomic classification for plants uses division not phylum

ALL plants . . .
- are multicellular
- are immobile
- have eukaryotic cells
- have cell walls made from cellulose
- are photosynthetic autotrophs
- reproduce using seeds or spores

There are four main Divisions of plants.

1. Division Bryophyta = mosses
   - use spores
   - no vascular tissue
   - terrestrial, live in moist places
   - small
   - do not have true stems, leaves, roots
   - occur in groups

2. Division Pterophyta = ferns
   - use spores
   - have vascular tissue
   - have true stems, leaves, roots

3. Division Gymnospermae
   - use seeds arranged on cones
   - use wind for pollination
   - have vascular tissue

4. Division Angiospermae
   - use seeds contained in flowers
   - have flowers that contain their reproductive organs
   - use animal pollinators
   - developing seeds are in the fruit
   - have vascular tissue

Division Angiospermae has two main classes:

1. Monocotyledons = monocots
   - grasses, corn, palm trees, bamboo, lilies
   - seeds are in one part
   - petals = multiples of 3
   - long, parallel-veined leaves

2. Dicotyledons = dicots
   - most angiosperms
   - seeds are in two parts
   - petals = multiples of 4 or 5
   - net-veined leaves

Seeds versus spores:

1. Seeds have food source with embryo—spores do not
2. Seeds can be transported longer distances
3. The outside covering of seeds is more protective than that of spores

WEEKLY SCHEDULE

Two Days

Day 1
- ❑ Lesson
- ❑ Lab/MSLab

Day 2
- ❑ FSS
- ❑ Lesson Review
- ❑ SWYK

Three Days

Day 1
- ❑ Lesson part 1
- ❑ Lab/MSLab–begin

Day 2
- ❑ Lesson part 2
- ❑ Lab/MSLab–complete

Day 3
- ❑ FSS
- ❑ Lesson Review
- ❑ SWYK

Five Days

Day 1
- ❑ Lesson part 1
- ❑ Lab/MSLab–begin

Day 2
- ❑ Lesson part 2
- ❑ Lab/MSLab–complete

Day 3
- ❑ FSS

Day 4
- ❑ Lesson Review

Day 5
- ❑ SWYK

Introduction

Note: Lab 32, in the next chapter, takes 5 to 10 days to see observable results. You might want to start it on Friday of the week you complete Chapter 31, or on Monday of the week you start Chapter 32.

Most of the species of organisms in domain Eukarya are in kingdom Animalia, the subject of Chapter 31. That makes for a long lesson listing traits. If you are using the two-days-a-week schedule, make sure your student stays focused for the entire lesson. The three- and five-day schedule split the lesson up over two days, studying invertebrates the first day and vertebrates the second.

The lab in this chapter, Arthropod Arrangement, has students comparing and discovering insects and arachnids. This is a general lab and a microscope lab combined. If you are not using a microscope for this course, this lab could be completed using only a magnifying glass (hand lens), but obviously students won't see as much detail for comparison.

Learning Goals

- Learn the traits all animals share.
- Learn the names of the nine major animal phyla.
- Learn shared derived traits that warrants an organism's placement in a phylum.
- Learn the difference between invertebrate and vertebrate animals.
- Learn the terms *ectotherm* and *endotherm*.
- Learn about the three types of mammals.

Extracurricular Resources

Books

You can find many books about animals at your local library. Some that I like are:

Animal Grossology, The Science of Creatures Gross and Disgusting, Branzei, Sylvia

Intelligence in Animals, Reader's Digest, long but very interesting

Time-Life Student Library Mammals

Wolf Pack: Tracking Wolves in the Wild, Johnson, Sylvia A. and Aamodt, Alice

Online

Visit Pandia Weblinks for videos and websites recommended for this chapter:

www.pandiapress.com/weblinks-biology2

Pandia PRESS

Lesson

Lords of Their Domain

You might have noticed that the lists of traits for each phylum only include those that are observable. There is no discussion about genetics, biochemistry, or biochemical pathways. There are differences in genetics and biochemistry between different organisms in the different phyla; I just do not discuss them. In fact, there must be genetic differences for groups with different traits—that is a given. There are certain genetic markers and biochemistry unique to each grouping. A discussion of the differences in chemistry and genetics is more appropriate for a high school or college-level course. That makes the written approach in Chapters 30, 31, and 32 more "old fashioned." As the genotypes of more and more organisms are known, there have been and will continue to be organisms that are reclassified based on the results of the analyses.

Lab and Microscope Lab

Arthropod Arrangement

When I wrote this lab I could just imagine being a scientist one hundred years ago, performing an experiment like this, just me and my microscope figuring out the difference between arachnids and hexapods. You need relatively fresh specimens. Insects that have been dead a long time tend to get crumbly. If you collect and kill live specimens, do not damage them. If you use bug spray on them be very careful. Bug spray is toxic to humans. You could collect them and put them in the freezer until they die. It all seems very unkind, doesn't it? I started looking for them for about two weeks leading up to the experiment; I found more than enough good specimens of fresh but dead insects. Spiders were harder to find, but I managed to find one too.

Possible Answers

This information is for a fly and a spider.

My Insect, Class Hexapod. My Observations:

Exoskeleton: *Hair and pores all over. I could see through the exoskeleton into the internal cavities.*

Body parts: *3 segments. I could see where the body parts are connected at joints, these occur at tapered sites where the joints connect them.*

Joints and legs: *6 legs. It is easy to find the points where the exoskeleton has its joints. Some joints look similar to how our joints would look if the skin were removed. Other joints look just like the joints I have seen on the exoskeleton of crab legs. It looks like the smaller of the two parts that connect the joint fits inside of the larger part.*

Other observations: *The fly had white lines across its abdomen that looked like segments in the exoskeleton. These were not joints, though. I wonder if the fly's*

exoskeleton grows in segments. Maybe every time the fly molts, its exoskeleton grows back with another segment. The fly looked like it had eggs in its abdomen. It was filled with little circles.

My Arachnid, Class Arachnida. My Observations:

Exoskeleton: *Hair and pores all over. I could see through the exoskeleton into the internal cavities.*

Body parts: *2 segments. I could see where the body parts are connected at joints, these occur at tapered sites where the joints connect them.*

Joints and legs: *8 legs. It is easy to find the points where the exoskeleton has its joints. Some joints look similar to how our joints would look if the skin were removed. Other joints look just like the joints I have seen on the exoskeleton of crab legs. It looks like the smaller of the two parts that connect by joint fits inside of the larger part. Looks like claws at the end of the legs.*

Other observations: *Some spiders, like the one I looked at, spin webs and have a little place at their bottom where the web-building material or silk comes from, called spinners. Insects do not have this. The spider has fangs, which look quite gruesome under the microscope. On either side of the fangs is what looks like teeth. These are not fangs or teeth such as mammals have. The spider's fangs and "teeth" look like they are a part of the exoskeleton.*

Similarities and Differences of Hexapods and Arachnids

Similarities:

An exoskeleton—The exoskeleton of both have hair and pores all over them. I could see through the exoskeleton on both into their internal cavities.

Jointed legs—It is easy to find the points where the exoskeleton has its joints. Both have claws at the ends of their legs.

Multiple body parts—You can see where the body parts are connected at joints, these occur at tapered sites where the joints connect them.

Differences:

Insects have 3 segments and 6 legs

Spiders have 2 segments and 8 legs

Some insects, like the one I looked at, have wings. Spiders do not have wings.

Spiders may have spinners or fangs. Insects do not.

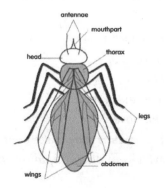

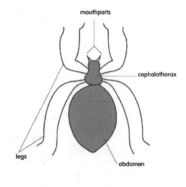

Famous Science Series

John James Audubon

John James Audubon is a famous naturalist and frontiersman. His story is interesting and unique. One reason I chose him was to give insight into how the methods of learning about animals has changed since his time.

John James Audubon painted hundreds of birds. Where was he born? When was he born? In what country did he study birds? Why did he go to that country? *He was born in 1785 in Haiti. He studied birds in the United States. He came to the U.S. to avoid being conscripted into Napoleon's army.*

What was the name of the book Audubon wrote about birds? *The Birds of America* or *Ornithological Biographies*

What society is named after him? What is its purpose? *The National Audubon Society. Its purpose is to save and restore ecosystems with special emphasis on the birds and wildlife found in those ecosystems.*

When Audubon discovered a bird he had not seen before, he shot it to study it more closely. Then he painted it. Do you think members of the National Audubon Society still shoot birds to study them? *Only with a camera!*

Show What You Know

Kingdom Animalia

1. This is the classification for the Southern Hairy-Nosed Wombat. Wombats are marsupials native to Australia and the island of Tasmania.

Domain	Eukarya
Kingdom	Animalia
Phylum	Chordata
Class	Mammalia
Order	Diprotodontia
Family	Vombatidae
Genus	*Lasiorhinus*
Species	*latifrons*

What is the scientific name for the Southern Hairy-Nosed Wombat? *Lasiorhinus latifrons*

2. Fix each statement to make it true. All animals...
   - are ~~unicellular~~ *multicellular*
   - are ~~immobile~~ *mobile*
   - have ~~prokaryotic~~ cells *eukaryotic*
   - ~~have~~ cell walls *do not have*
   - ~~are autotrophs, which means they make their own food using photosynthesis~~ *are heterotrophs, which eat other organisms for food*

3. Match each group with the best description for it.

Crustacea — long bodies, lots of segments, 1 to 2 feet coming from each segment

Myriapoda — 2 body segments, 8 legs

Hexapoda — 3 body segments, 6 legs, 2 antennae

Arachnida — 10 to 40 legs, 4 antennae, gills

4. Fill in the cladogram.

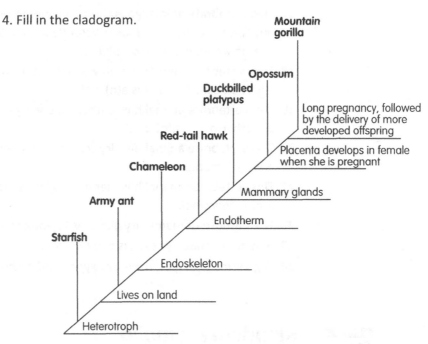

Multiple Choice

1. A squid is an invertebrate animal, meaning *it does not have a backbone.*
2. An animal that has one muscular foot and a soft body is a(n) *mollusk.*
3. An animal that is aquatic, does not have tissue but does have specialized cells, has a hollow body with pores in it, and has a big hole on top where waste flows out, is a(n) *porifera.*
4. An animal with jointed legs, a segmented body, and an exoskeleton is a(n) *arthropod.*
5. An aquatic animal with a radial body plan, a sac-like body with one opening, and stinging tentacles it uses to immobilize it prey is a(n) *cnidaria.*
6. An aquatic animal with a radial body plan, a tough spiny skin, and tube feet it uses to move is a(n) *echinoderm.*
7. An animal with a backbone, a head, and a sophisticated body plan is a(n) *chordate.*
8. Segmented worms *have both male and female parts.*
9. Nematodes are roundworms that *have a long, threadlike body.*
10. Platyheminthes are worms that *have a flat body with a mouth at one end.*
11. This animal gets food when it moves food and water through pores in its body. It is a *sponge.*

12. The term *radial body plan* means an organism *has a central point that the rest of their body is arranged around.*

13. A vertebrate has *an internal skeleton.*

14. Endoskeletons are made from *bone and cartilage.*

15. An ectotherm *regulates its body temperature by exchanging heat with the environment.*

16. An endotherm *regulates its own body temperature internally.*

17. This vertebrate animal goes through a metamorphosis, where it starts out as one form and grows to look differently as an adult, is an ectotherm, and lays eggs in water. It is a(n) *amphibian.*

18. This vertebrate animal lives in water, has fins, breathes through gills, lays eggs, and is an ectotherm. It is a(n) *fish.*

19. This vertebrate animal has feathers and wings, a beak, lays eggs, and is an endotherm. It is a(n) *bird.*

20. This vertebrate animal has dry scaly skin, lays eggs, and is an ectotherm. It is a(n) *reptile.*

21. This vertebrate animal has mammary glands, hair or fur, and is an endotherm. It is a(n) *mammal.*

22. The purpose of mammary glands is to *make milk.*

23. A mammal that lays eggs is a *monotreme.*

24. Organisms with gills *transfer oxygen and carbon dioxide across them.*

Lesson Review

Kingdom Animalia

Try using the Socratic approach and see how well students remember the traits with or without a little prompting.

All animals
- Are multicellular
- Are mobile
- Have eukaryotic cells
- Do not have cell walls
- Are heterotrophs

Invertebrate = no backbone
Vertebrate = has a backbone

Eight of the nine major animal phyla are invertebrates:

1. **Porifera = sponges**
   - Aquatic
   - Do not have tissues or organs
   - Do have specialized cells
   - Hollow body
   - Food and water move into the sponge's body through openings called pores
   - Waste moves out of the sponge's body through a hole in its top
   - Hard spine that gives them support and shape

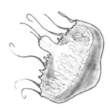

2. Cnidaria = jellyfish, sea anemones
- Aquatic
- Radial body plan
- Sac-like body with one opening, their mouth
- Tentacles with stinging cells surrounding their mouth

3. Platyhelminthes = flat worm

- Flat body
- Mouth at one end
- Parasites

4. Nematoda = roundworm
- Long, threadlike bodies
- Parasites

5. Annelida = earthworm
- Tube-shaped bodies made from segments
- Both male and female parts on same animal

6. Mollusca = shelled animals
- Live in water, except snails, slugs
- Shell, except squids, octopuses, some species of slugs
- Muscular foot that they use so they can move and burrow
- Soft bodies with a layer of folded skin that protects their internal organs

7. Arthropoda = insects, spiders
- Jointed legs
- Bodies are divided into segments
- Exoskeleton
- Antennae, except arachnids
- Four main classes = **Crustacea, Myriapoda, Hexapoda, Arachnida**

8. Echinoderms = starfish, sand dollars
- Live in the marine biome
- Tough, spiny skin
- Radial body plan arranged in five parts
- Move on tube feet by creating and releasing suction

One animal phylum contains all vertebrates:
9. Chordata = animals with backbones
- Endoskeleton with backbone
- Head
- Sophisticated body system

Classes of Chordates:

Fish

- Live in water
- Bullet, streamlined shape
- Fins for swimming
- Breathe through gills
- Most ectotherms
- Most lay eggs

Amphibians

- Lay eggs in water
- Go through metamorphosis
- Smooth moist skin they exchange oxygen across
- Ectotherms

Reptiles

- Dry, scaly skin
- Breathe through lungs
- Lay eggs that have a leathery shell
- When they hatch from their egg they look like miniature adults
- Ectotherms

Birds

- Feathers and wings
- Two legs covered in scaly skin
- Beak with no teeth
- Breathe through lungs
- Lay eggs in hard shells
- Endotherms
- Most can fly

Mammals

- Hair or fur
- Breathe through lungs
- Mammary glands
- Endotherms

Three groups of mammals with three different birth strategies:

1. Placental mammals, e.g. wildebeest
   - The embryo attached to a placenta
   - Nutrients and waste are transferred across the placenta between mother and embryo
   - Fetus born well-developed

2. Marsupial mammals, e.g. kangaroo
   - The embryo attached to a placenta
   - Nutrients and waste are transferred across the placenta between mother and embryo
   - Fetus born less developed

3. Monotreme mammals, e.g. duckbilled platypus
   - Lay eggs
   - Embryo develops in egg
   - No placenta

Unit VII: Classification
Chapter 32: Kingdoms Fungi and Protists

WEEKLY SCHEDULE

Two Days

Day 1
❑ Lesson
❑ Lab
❑ Dissection/MSLab

Day 2
❑ FSS
❑ Lesson Review
❑ SWYK
❑ Unit VII Exam

Three Days

Day 1
❑ Lesson
❑ Lab

Day 2
❑ Dissection/MSLab

Day 3
❑ FSS
❑ Lesson Review
❑ SWYK
❑ Unit VII Exam

Five Days

Day 1
❑ Lesson

Day 2
❑ Lab

Day 3
❑ Dissection/MSLab

Day 4
❑ FSS

Day 5
❑ Lesson Review
❑ SWYK
❑ Unit VII Exam

FSS: Famous Science Series
MSLab: Microscope Lab
SWYK: Show What You Know

Introduction

Lab 32 takes 5 to 10 days to see observable results. Start it on Friday of the week you do Chapter 31 or on Monday of the week you start Chapter 32.

Chapter 32 explains the last two kingdoms, both of which are in domain Eukarya. The two kingdoms are kingdom Fungi, which is filled with chemotrophs, and kingdom Protista, which has all the eukaryotes that do not fit into the other three kingdoms in domain Eukarya. Both labs and the Famous Science Series in Chapter 32 deal with organisms in kingdom Fungi. There is so much diversity within kingdom Protista I did not feel that focusing on one member of that kingdom gave much information from an overall perspective. Instead I felt a focus on fungi would lead to a deeper and more meaningful understanding of kingdom Fungi as a whole.

This is the last chapter in Unit VII. There is a Unit VII Exam that covers the material found in chapters 28 through 32, in the appendix of the student Workbook. The answer key is found at the end of this chapter.

Learning Goals

- Learn traits all fungi share.
- Learn the parts of a typical fungus.
- Learn about chemotrophy as a feeding strategy.
- Recognize issues surrounding many of the classifications of organisms in kingdom Protista.
- Learn the traits that define and separate the three types of protists.
- Learn about the symbiotic relationship that makes lichens.

Extracurricular Resources

Books
Protists: Algae, Amoebas, Plankton, and Other Protists, Arato, Rona
Fungi: Mushrooms, Toadstools, Molds, Yeasts, and Other Fungi, Wearing, Judy
Truffle Trouble: The Case of the Fungus Among Us, Lloyd, Emily
The KidHaven Science Library - Molds and Fungi, Silverman, Buffy

Online
Visit Pandia Weblinks for videos and websites recommended for this chapter:

www.pandiapress.com/weblinks-biology2

Lesson

Fungus Like Dung, Protists Like Water

Some members of kingdom Protista are more closely related to plants and others are more closely related to fungi and animals. The pairing of the two kingdoms in this chapter does not mean to imply a closer relationship than there is between them.

Lab

Watching Fungi Feed

The riper the banana, the faster this experiment will go. Make sure you have a baggie with no holes because you are going to end up with a rather gross and soupy mess. Plus, you do not want the carbon dioxide to leak out of it. This experiment is a good place to review cellular respiration and its by-products.

Possible Answers

Hypothesis: *I think the banana without fungi will rot more but the banana with fungi will decompose faster.*

Observations

Day 1: *The banana with fungi looks like a candy because of the yeast all over it looking like sprinkles. The banana without yeast looks like a normal, just-peeled banana.*

Day 2: *Both bananas look the same as they did yesterday.*

Day 3: *The banana with yeast is starting to get goopy, there is gas in the bag, and the banana is dissolving away. The banana without yeast is starting to brown slightly.*

Day 4: *The banana with yeast: the bag is almost full of gas, the banana is very goopy and smaller, and it is rotting and decomposing much faster than the other banana. The banana without yeast is a little browner. No gas is accumulating in that baggie.*

Day 5: *The banana with yeast: the banana has turned into an unrecognizable mush. The bag is full of CO_2 gas. The banana without yeast is a little browner every day. This banana did not rot much and did not decompose. No gas is accumulating in the baggie.*

Results/Conclusion: *The yeast broke down the glucose molecules in the banana to make food for themselves. You could tell this because of the CO_2 gas that filled the baggie. As the yeast ate the banana, the banana decomposed and rotted away. The control banana, the one with no yeast in the baggie, did not decompose and only turned a little brown. No gas accumulated in the baggie; from this I conclude CO_2 gas was not being produced.*

Bonus question: What reaction do you think is occurring if carbon dioxide is accumulating inside the baggie? Why is it accumulating and not cycling? What would happen if the gas inside the baggie ran out of oxygen? *The reaction is cellular respiration if CO_2 gas is being produced. It is not cycling because there are no photosynthesizing plants in the baggies to use up the CO_2 gas. If the oxygen was used up, another type of respiration, fermentation, would start to occur.*

Dissection and Microscope Lab

Fungus Up Close and Personal

Have students take their time with the dissection and slide preparation. Remind them to keep the slices thin and be careful with the mushroom. Mushrooms are delicate. The lab sheet ends with a question about whether chloroplasts are observed. They won't be. Make sure students understand that there are no chloroplasts because mushrooms and other fungi are not plants or plantlike protists and do not photosynthesize. It is a common misconception that mushrooms are plants.

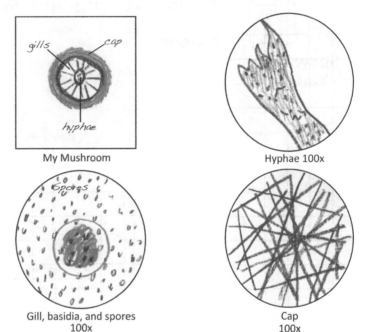

My Mushroom

Hyphae 100x

Gill, basidia, and spores
100x

Cap
100x

Did you see any chloroplasts? Why or why not? *No, mushrooms are not plants and do not use photosynthesis for making food and therefore have no chloroplasts. Or: Mushrooms are chemotrophs, not autotrophs; therefore they do not have chloroplasts.*

Famous Science Series

Truffle Pigs

What are truffles, what are they used for, and what is the job of a truffle pig? *Truffles are a type of fungus that people use in gourmet cooking. The job of a truffle pig is to find truffles.*

Where do truffles grow? *Underground*

When did people start using pigs to locate truffles? *The first documented use was from the 15th century, but historians think pigs have been used to hunt truffles since Roman times.*

Another animal is replacing the truffle pig. What kind of animal is it and why is the pig being replaced? *Truffle pigs are being replaced by truffle hounds, dogs that hunt for and locate truffles. The dogs are better at giving the truffles to their masters Instead of eating the truffles, which can be a problem with pigs.*

Show What You Know

Kingdoms Fungi and Protists

Cladogram

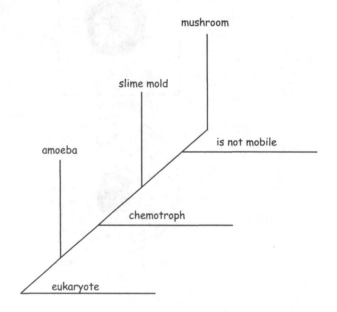

Classification

Domain	*Eukarya*
Kingdom	*Fungi*
Phylum	Ascomycota
Class	Pezizomycetes
Order	Pezizales
Family	Tuberaceae
Genus	*Tuber*
Species	*borchii*

What is the scientific name of this truffle? *Tuber borchii*

Multiple Choice

1. Algae are *plantlike protists.*
2. Amoebas are *animal-like protists.*
3. Mushrooms are *fungi.*
4. Slime molds are *fungi-like protists.*
5. Plantlike protists are *autotrophs.*
6. Animal-like protists are *heterotrophs.*
7. Fungi-like protists are *All of the above*
8. Which characteristic is NOT shared between fungi and fungi-like protists? *They are mobile.*
9. Which characteristic is NOT shared between animals and animal-like protists? *They are unicellular.*
10. Which trait is NOT shared between plants and plant-like protists? *They are mobile.*
11. Protists *All of the above*
12. Chemotrophs *secrete chemicals that break material into molecules they can absorb.*
13. Lichens are a symbiotic relationship between what two types of organisms? *Fungi and algae*
14. Fungi reproduce *using spores.*
15. Hyphae *are used for absorbing nutrients.*

Lesson Review

Kingdoms Fungi and Protists

Try using the Socratic approach and see how well students remember the traits with or without a little prompting.

Fungi are decomposers. Decomposers are consumers (heterotrophs) that recycle nutrients in the community. They are important to the nitrogen cycle and the phosphorus cycle.

Shared Derived Traits of Fungi
- Eukaryotic cells
- Multicellular, except yeast
- Use spores to reproduce
- Have cell walls made from chitin
- Are chemotrophs
- Are not mobile

Chemotrophs = feed by secreting chemicals that turn food into molecules they can absorb

Mushrooms use a structure called a hyphae to secrete the chemicals, absorb food molecules, and transport them

Protists are a diverse group of organisms that live in watery environments. There are three types:

1. Plantlike protists
   - Photosynthetic autotrophs
   - Eukaryotes
   - Unicellular and multicellular
   - Are unlike plants in that they have no specialized tissue and are immobile

2. Animal-like protists
   - Heterotrophs
   - Eukaryotes
   - Mobile
   - Are unlike animals in that they are unicellular

3. Fungi-like protists
   - Chemotrophs
   - Eukaryotes
   - Cell walls
   - Reproduce using spores
   - Are unlike fungi in that they are mobile

Unit VII: Classification

Answer Key Unit Exam Chapters 28–32

The exam for Unit VII is found in the appendix of the student Workbook.

1. **Multiple Choice.** Each of the following multiple choice questions gives one or more traits. Choose the organism that is described by the trait(s). (2 points each, 40 points total)

I am a prokaryote who is classified based on differences in my RNA. What am I? *Archaea*

I am a multicellular, eukaryotic autotroph. What am I? *Plantae*

I am a eukaryotic chemotroph. I am immobile. What am I? *Fungi*

I am a unicellular prokaryote who is classified by shape, feeding strategy, and the conditions in which I grow. What am I? *Bacteria*

I am a multicellular, eukaryotic heterotroph. I do not have cell walls. What am I? *Animalia*

I am a eukaryote that lives in watery places. I am usually unicellular. What am I? *Protista*

I am in kingdom Plantae. I am small with no vascular tissue. I reproduce using spores. What am I? *Bryophyte*

I am in kingdom Plantae. I have vascular tissue and reproduce with seeds on cones. What am I? *Gymnosperm*

I am in kingdom Plantae. I have vascular tissue and reproduce with seeds contained in flowers. What am I? *Angiosperm*

I am in kingdom Plantae. I have vascular tissue and reproduce with spores. What am I? *Pterophyte*

I am in kingdom Animalia. I have a backbone. What am I? *Chordata*

I am in kingdom Animalia. I have a shell and a muscular foot I use to move. What am I? *Mollusca*

I am in kingdom Animalia. I have jointed legs and an exoskeleton. What am I? *Arthropoda*

I am in kingdom Animalia. I am a parasitic worm. I have a long, threadlike body. What am I? *Nematoda*

I am in kingdom Animalia. I am a sponge. I have pores all over my body. What am I? *Porifera*

I am in kingdom Animalia. I am an earthworm. I have a segmented body with male and female parts on it. What am I? *Annelida*

I am in kingdom Animalia. I live in the marine biome. I have tough, spiny skin and a radial body with five sections. What am I? *Echinodermata*

I am in kingdom Animalia. I am a jellyfish. I have tentacles with stinging cells around my mouth that I use to paralyze my prey. What am I? *Cnidaria*

I am in kingdom Animalia. I am a parasitic worm with a flat body. What am I? *Platyhelminthes*

The three types of mammals are *monotreme, placental, marsupial.*

2. **Matching.** Phylum Chordata has five classes. Match the class with the description. (1 point each, 5 total)

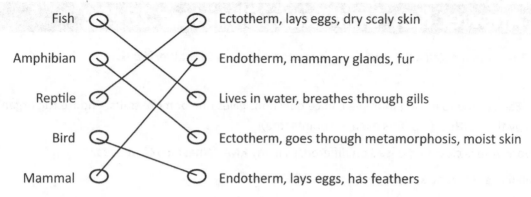

Fish	Ectotherm, lays eggs, dry scaly skin
Amphibian	Endotherm, mammary glands, fur
Reptile	Lives in water, breathes through gills
Bird	Ectotherm, goes through metamorphosis, moist skin
Mammal	Endotherm, lays eggs, has feathers

3. **Short Answers**

   What evidence is used to classify organisms? (2 points each, 8 total)

   Physical appearance

   Method of getting nutrients; autotroph or heterotroph

   Anatomy

   Genetic evidence (or cell structure)

   What are the three domains called? (2 points each, 6 total)

   Bacteria, Archaea, Eukarya

4. **Vocabulary match.** Match the word with the definition that best fits. (1 point each, 10 total)

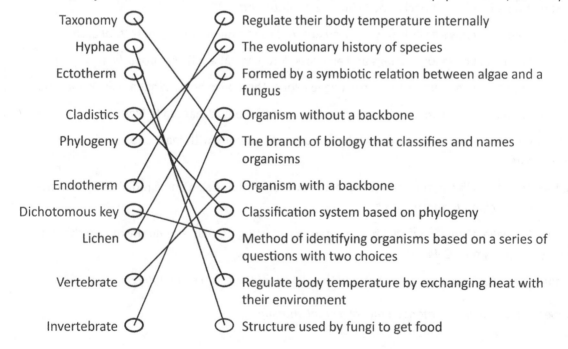

Taxonomy	Regulate their body temperature internally
Hyphae	The evolutionary history of species
Ectotherm	Formed by a symbiotic relation between algae and a fungus
Cladistics	Organism without a backbone
Phylogeny	The branch of biology that classifies and names organisms
Endotherm	Organism with a backbone
Dichotomous key	Classification system based on phylogeny
Lichen	Method of identifying organisms based on a series of questions with two choices
Vertebrate	Regulate body temperature by exchanging heat with their environment
Invertebrate	Structure used by fungi to get food

5. **Classification.** On the right side is the classification for a vampire squid. Fill in the left side of the table with the names of the eight levels of classification. (1 point each, 8 points total)

Domain	Eukarya
Kingdom	Animalia
Phylum	Mollusca
Class	Cephalopoda
Order	Vampyromorphida
Family	Vampyroteuthidae
Genus	Vampyroteuthis
Species	infernalis

What is the scientific name of the vampire squid? What is this system of naming organisms called? (1 point correct name, 1 for correct punctuation, 1 for binomial classification, 3 total) *Vampyroteuthis infernalis, binomial classification*

6. **Cladogram** (2 points each, 20 points total)

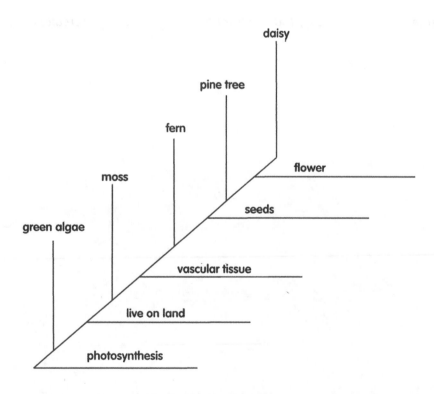

7. Extra Credit

There are four main classes of arthropods. What are they called? (2 points for each name, 8 total)
Name 1 organism from each class. (1 point for each name, 4 total)
What is 1 characteristic specific to each class? (1 point for each trait, 4 total)
  (16 possible extra credit points)

1. *Crustacea*
   • *1 of these–lobsters, crabs, and shrimp.*
   • *1 of these–between 10 to 40 legs, 2 pairs of antennae, breathe through gills*
2. *Myriapoda*
   • *1 of these–millipede, centipede*
   • *1 of these–long bodies made from lots of segments, centipedes have 1 pair of legs coming from each segment, millipedes have 2 pairs of legs coming from each segment*
3. *Hexapoda*
   • *Any insect will do*
   • *1 of these–3 body divisions (head, thorax, abdomen), 6 legs, 1 pair of antennae, 1 or 2 pairs of wings*
4. *Arachnida*
   • *1 of these–spider, scorpion, mite, tick*
   • *1 of these–2 body divisions, 8 legs, no antennae, no wings*

Point total: */100, 100 points for the exam and 16 extra credit points.*